Everyday Mathematics®

The University of Chicago School Mathematics Project

Study Links

Grade 4

Mc Graw Hill **Wright Group**

The McGraw·Hill Companies

The University of Chicago School Mathematics Project (UCSMP)

Max Bell, Director, UCSMP Elementary Materials Component; Director, *Everyday Mathematics*
First Edition
James McBride, Director, *Everyday Mathematics* Second Edition
Andy Isaacs, Director, *Everyday Mathematics* Third Edition
Amy Dillard, Associate Director, *Everyday Mathematics* Third Edition

Authors

Max Bell, John Bretzlauf, Amy Dillard, Robert Hartfield, Andy Isaacs, James McBride,
Kathleen Pitvorec, Peter Saecker, Robert Balfanz*, William Carroll*, Sheila Sconiers*

*First Edition only

Technical Art	**Teachers in Residence**	**Editorial Assistant**
Diana Barrie	Carla L. LaRochelle	Laurie K. Thrasher
	Rebecca W. Maxcy	

Contributors

Martha Ayala, Virginia J. Bates, Randee Blair, Donna R. Clay, Vanessa Day, Jean Faszholz,
James Flanders, Patti Haney, Margaret Phillips Holm, Nancy Kay Hubert, Sybil Johnson,
Judith Kiehm, Deborah Arron Leslie, Laura Ann Luczak, Mary O'Boyle, William D. Pattison,
Beverly Pilchman, Denise Porter, Judith Ann Robb, Mary Seymour, Laura A. Sunseri-Driscoll

Photo Credits

©Gregory Adams/Getty Images, cover, top right; ©Getty Images, cover, center, pp. iv, 41; ©Tony
Hamblin; Frank Lane Picture Agency/Corbis, cover, cover, bottom left.

 This material is based upon work supported by the National Science Foundation under Grant
No. ESI-9252984. Any opinions, findings, conclusions, or recommendations expressed in this
material are those of the authors and do not necessarily reflect the views of the National
Science Foundation.

---

www.WrightGroup.com

Printed in the United States of America.

Send all inquiries to:
Wright Group/McGraw-Hill
P.O. Box 812960
Chicago, IL 60681

ISBN-13 978-0-07-609741-8
ISBN-10 0-07-609741-2

5 6 7 8 9 DBH 12 11 10 09 08 07

Contents

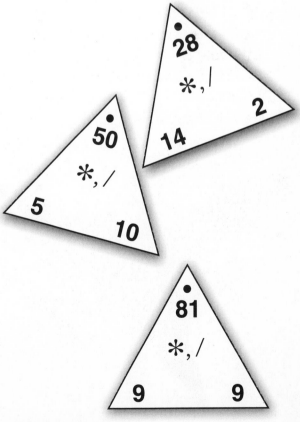

Contents   **v**

STUDY LINK 1·1

Unit 1: Family Letter

Introduction to *Fourth Grade Everyday Mathematics*®

Welcome to *Fourth Grade Everyday Mathematics.* It is part of an elementary school mathematics curriculum developed by the University of Chicago School Mathematics Project (UCSMP).

Everyday Mathematics offers students a broad background in mathematics. Some approaches may differ from those you used as a student, but the approaches used are based on research, field test results, and the mathematics students will need in this century.

Fourth Grade Everyday Mathematics emphasizes the following content:

Algebra and Uses of Variables Reading, writing, and solving number sentences

Algorithms and Procedures Exploring addition, subtraction, multiplication, and division methods; inventing individual procedures and algorithms; and experimenting with calculator procedures

Coordinate Systems and Other Reference Frames Using numbers in reference frames: number lines, coordinates, times, dates, and latitude and longitude

Exploring Data Collecting, organizing, displaying, and interpreting numerical data

Functions, Patterns, and Sequences Designing, exploring, and using geometric and number patterns

Geometry and Spatial Sense Developing an intuitive sense about 2- and 3-dimensional objects, their properties, uses, and relationships

Measures and Measurement Exploring metric and U.S. customary measures: linear, area, volume, weight; and exploring geographical measures

Numbers, Numeration, and Order Relations Reading, writing, and using whole numbers, fractions, decimals, percents, negative numbers; and exploring scientific notation

Operations, Number Facts, and Number Systems Practicing addition and subtraction to proficiency; and developing multiplication and division skills

Problem Solving and Mathematical Modeling Investigating methods for solving problems using mathematics in everyday situations

Naming and Constructing Geometric Figures

During the next few weeks, the class will study the geometry of 2-dimensional shapes. Students will examine definitions and properties of shapes and the relationships among them. Students will use compasses to construct shapes and to create their own geometric designs.

Please keep this Family Letter for reference as your child works through Unit 1.

Vocabulary

Important terms in Unit 1:

concave (nonconvex) polygon A polygon in which at least one vertex is "pushed in."

concave polygon

convex polygon A polygon in which all vertices are "pushed outward."

convex polygon

endpoint A point at the end of a line segment or a ray.

line Informally, a straight path that extends infinitely in opposite directions.

line segment A straight path joining two points. The two points are called the endpoints of the segment.

parallelogram A quadrilateral that has two pairs of parallel sides. Opposite sides of a parallelogram have equal lengths. Opposite angles of a parallelogram have the same measure.

polygon A 2-dimensional figure that is made up of three or more line segments joined end to end to make one closed path. The line segments of a polygon may not cross.

quadrangle (quadrilateral) A polygon that has four sides and four angles.

ray A straight path that extends infinitely from a point called its endpoint.

rhombus A quadrilateral whose sides are all the same length. All rhombuses are parallelograms. Every square is a rhombus, but not all rhombuses are squares.

trapezoid In *Everyday Mathematics,* a quadrilateral that has exactly one pair of parallel sides.

vertex The point where the rays of an angle, the sides of a polygon, or the edges of a polyhedron meet.

Do-Anytime Activities

To work with your child on concepts taught in this unit, try these interesting and rewarding activities:

1. Help your child discover everyday uses of geometry as found in art, architecture, jewelry, toys, and so on.

2. See how many words your child can think of that have Greek/Latin prefixes such as *tri-, quad-, penta-, hexa-,* and *octa-.*

3. Help your child think of different ways to draw or make figures without the use of a compass, protractor, or straightedge. For example, you can trace the bottom of a can to make a circle, bend a straw to form a triangle, or make different shapes with toothpicks.

4. Challenge your child to draw or build something, such as a toothpick bridge, using triangular and square shapes. Or show pictures of bridges and point out the triangles used in bridges to provide support.

Building Skills through Games

In Unit 1, your child will play the following games.

Addition Top-It See *Student Reference Book,* page 263. This game provides practice with addition facts.

Polygon Pair-Up See *Student Reference Book,* page 258. This game provides practice identifying properties of polygons.

Sprouts See *Student Reference Book,* page 313. This game provides practice with simple vertex-edge graphs and developing game strategies.

Subtraction Top-It See *Student Reference Book,* pages 263 and 264. This is a variation of *Addition Top-It* and provides practice with subtraction facts.

Sz'kwa See *Student Reference Book,* page 310. This game provides practice with intersecting line segments and developing game strategies.

As You Help Your Child with Homework

As your child brings assignments home, you may want to go over the instructions together, clarifying them as necessary. The answers listed below will guide you through this unit's Study Links.

Study Link 1·2

2. a.
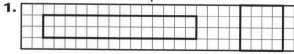

b.

c. The line has arrows on both ends, but the line segment does not.

3. a.

b. No. A ray's endpoint must be listed first when naming a ray.

4. A ruler has markings on it, so it can be used to measure.

Study Link 1·3

Sample answers:

1.
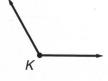

2.

3. The polygons in Problems 1 and 2 have 4 sides and at least 1 pair of parallel sides. The Problem 1 polygons have 2 pairs of equal, parallel sides and all right angles.

4. a.   b. E   c. FED

5.

Study Link 1·4

1. Sample answer:
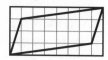

2. a. yes   b. yes   c. yes   d. no

3. Sample answer: 4. kite
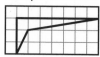 

Study Link 1·5

1. rectangle   2. Equilateral triangle

3. rhombus

Study Link 1·6

1. A, B, C, E, F, G, I   2. B, C

3. C, E, F, I   4. A

5. A, B, D, F, G, H, I   6. D, G, H

7. 2

Study Link 1·8

1. Sample answers:

a. square   c. hexagon

2. Sample answer: Sides are all the same length, and interior angles are all the same measure.

4

STUDY LINK 1·2 | Line Segments, Lines, and Rays

SRB
90 91

1. List at least 5 things in your home that remind you of line segments.

_____

_____

_____

Use a straightedge to complete Problems 2 and 3.

2. **a.** Draw and label line *AB*.

   b. Draw and label line segment *AB*.

   c. Explain how your drawings of $\overleftrightarrow{AB}$ and $\overline{AB}$ are different.

   _____

   _____

3. **a.** Draw and label ray *CD*.

   b. Anita says $\overrightarrow{CD}$ can also be called $\overrightarrow{DC}$. Do you agree? Explain.

   _____

   _____

4. Explain how a ruler is different from a straightedge.

   _____

   _____

| **Practice** |

5. 13 − 7 = _____

6. 15 − 8 = _____

7. _____ = 90 − 50

8. 140 − 60 = _____

9. _____ = 57 − 39

10. 115 − 86 = _____

5

STUDY LINK
1·3

Angles and Quadrangles

Use a straightedge to draw the geometric figures.

1. Draw 2 examples of a rectangle.

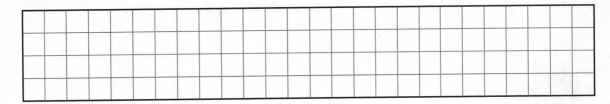

2. Draw 2 examples of a trapezoid.

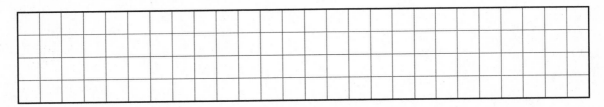

3. How are the polygons in Problems 1 and 2 similar? How are they different?

_____

_____

_____

4. a. Draw right angle *DEF*.

5. Draw an angle that is larger than a right angle. Label the vertex *K*.

b. What is the vertex of the angle? Point _____

c. What is another name for ∠*DEF*? ∠ _____

Practice

6. $9 + 8 =$ _____

7. $7 + 8 =$ _____

8. $30 + 80 =$ _____

9. _____ $= 50 + 40$

10. _____ $= 17 + 94$

11. $158 + 93 =$ _____

7

STUDY LINK 1·4 Classifying Quadrangles

1. A parallelogram is a quadrangle (quadrilateral) that has 2 pairs of parallel sides.

Draw a parallelogram.

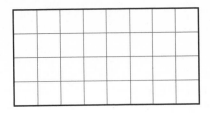

2. Answer *yes* or *no*. Explain your answer.

a. Is a rectangle a parallelogram? _____

_____

b. Is a square a parallelogram? _____

_____

c. Is a square a rhombus? _____

_____

d. Is a trapezoid a parallelogram? _____

_____

3. Draw a quadrangle that has at least 1 right angle.

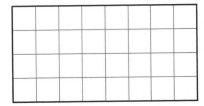

4. Draw a quadrangle that has 2 pairs of equal sides but is NOT a parallelogram.

This is called a _____.

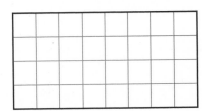

| **Practice** |

5. $12 - 6 =$ _____

6. $16 - 7 =$ _____

7. $210 - 150 =$ _____

8. _____ $= 140 - 80$

9. _____ $= 93 - 58$

10. $123 - 76 =$ _____

STUDY LINK 1·5 **Polygon Riddles**

Answer each riddle. Then use a straightedge to draw a picture
of the shape in the space to the right.

1. I am a quadrangle.
I have 2 pairs of parallel sides.
All of my angles are right angles.
I am not a square.

What am I? _____

2. I am a polygon.
All of my sides have the same measure.
All of my angles have the same measure.
I have 3 sides.

What am I? _____

3. I am a polygon.
I am a quadrangle.
All of my sides are the same length.
None of my angles are right angles.

What am I? _____

Try This

4. On the back of this page, make up your own polygon riddle using 4 clues.
Make 2 of the clues hard and 2 of the clues easy. Check your riddle by using a
straightedge to draw a picture of the polygon. Ask a friend or someone at
home to solve your polygon riddle.

Practice

5. 8 + 9 = _____ **6.** 7 + 8 = _____ **7.** 90 + 70 = _____

8. _____ = 60 + 50 **9.** _____ = 54 + 59 **10.** 185 + 366 = _____

STUDY LINK 1·6 Properties of Geometric Figures

SRB
96–100

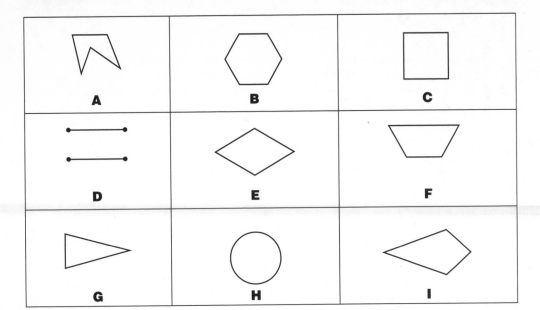

Write the letter or letters that match each statement.

1. These are polygons. _____

2. These are regular polygons. _____

3. These are quadrangles. _____

4. These are concave. _____

5. These are NOT parallelograms. _____

6. These do NOT have any right angles or angles whose measures are larger than a right angle. _____

Try This

7. Take a paper clip and two pencils. Create a homemade compass. You may not bend or break the paper clip. How many different size circles can you make with it? _____

Practice

8. 30 + 50 = _____   **9.** 40 + 60 = _____   **10.** 250 + 140 = _____

11. _____ = 80 − 20   **12.** _____ = 120 − 70   **13.** 460 − 230 = _____

13

STUDY LINK
1·7 | # The Radius of a Circle

1. Find 3 circular objects. Trace around them to make 3 circles in the space below or on the back of this page. For each circle, do the following:

SRB
104

a. Draw a point to mark the approximate center of the circle. Then draw a point on the circle.

b. Use a straightedge to connect these points. This line segment is a **radius** of the circle.

Example:

c. Use a ruler to measure the radius to the nearest centimeter. If you do not have a ruler at home, cut out the one at the bottom of this page.

d. Record the measure of the radius next to the circle.

Practice

2. _____ = 80 + 20   **3.** _____ = 30 + 90   **4.** 580 + 370 = _____

5. 120 − 30 = _____   **6.** 160 − 70 = _____   **7.** 650 − 280 = _____

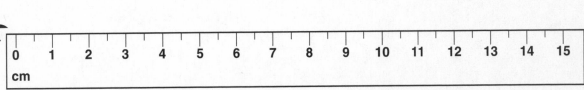

15

STUDY LINK
1·8 | **Inscribed Polygons**

1. Use a straightedge to inscribe a different polygon in each of the circles below. Write the name of each polygon.

Example:

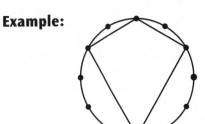

kite

a. _____

b. _____

c. _____

d. _____

2. Are any of the polygons that you drew *regular polygons?* Explain how you know.

_____

_____

_____

Practice

3. 41 + 27 = _____

4. _____ = 263 + 59

5. 461 + 398 = _____

6. _____ = 72 − 36

7. 158 − 71 = _____

8. 742 − 349 = _____

Unit 2: Family Letter

Using Numbers and Organizing Data

Your child is about to begin this year's work with numbers. The class will examine what numbers mean and how they are used in everyday life.

In today's world, numbers are all around us—in newspapers and magazines and on TV. We use them

◆ to count things *(How many people are in the room?)*

◆ to measure things *(How tall are you?)*

◆ to create codes *(What is your Social Security number?)*

◆ to locate things in reference frames *(What time is it?)*

◆ to express rates, scales, and percents *(How many miles per gallon does your car get? What percent voted for Jamie?)*

Sometimes students will need to interpret a collection of numbers. The class will learn to organize such collections of numbers in tables and graphs and to draw conclusions about them.

Computation is an important part of problem solving. Fortunately, we are no longer restricted to paper-and-pencil methods of computation. We can use calculators or computer programs to solve lengthy or complex problems. Your child will practice mental and paper-and-pencil methods of computation, use a calculator, and have opportunities to decide which is most appropriate for solving a particular problem.

Many of us were taught that there is just one way to do computations. For example, we may have learned to subtract by "borrowing." We may not have realized that there are other ways of subtracting numbers. While students will not be expected to learn more than one method, they will examine several different methods and realize that there are often several ways to arrive at the same result. They will have the option of using the methods with which they are comfortable or even inventing one of their own.

Mathematics games will be used throughout the school year to practice various arithmetic skills. Through games, practice becomes a thinking activity to be enjoyed. The games your child will play in this unit will provide practice with renaming numbers, with addition, and with subtraction. They require very little in the way of materials, so you may play them at home as well.

Please keep this Family Letter for reference as your child works through Unit 2.

STUDY LINK 1·9 | **Unit 2: Family Letter** *cont.*

Vocabulary

Important terms in Unit 2:

algorithm A set of step-by-step instructions for doing something, such as carrying out a computation or solving a problem.

base 10 Our number system in which each place in a number has a value 10 times the place to its right and $\frac{1}{10}$ the place to its left.

column-addition A method for adding numbers in which the addends' digits are first added in each place-value column separately, and then 10-for-1 trades are made until each column has only one digit. Lines are drawn to separate the place-value columns.

	100s	10s	1s
	2	4	8
+	1	8	7
Add the columns:	3	12	15
Adjust the 1s and 10s:	3	13	5
Adjust the 10s and 100s:	4	3	5

equivalent names Different names for the same number. For example, $2 + 6$, $4 + 4$, $12 - 4$, $18 - 10$, $100 - 92$, $5 + 1 + 2$, eight, VIII, and $\cancel{IIII}\ III$ are equivalent names for 8.

line plot A sketch of data in which check marks, Xs, or other marks above a labeled line show the frequency of each value.

```
                    X
                    X    X
                    X    X
Number          X   X    X
of Students     X   X    X          X
              ──┼───┼────┼────┼────┼──►
                0   1    2    3    4
                  Number of Siblings
```

mean The sum of a set of numbers divided by the number of numbers in the set. The mean is often referred to simply as the "average."

median The middle value in a set of data when the data are listed in order from least to greatest. If there is an even number of data points, the median is the *mean* of the two middle values.

mode The value or values that occur most often in a set of data.

name-collection box A diagram that is used for writing *equivalent names* for a number. The box to the right shows names for 8.

8
2 + 6
4 + 4
VIII
eight

partial-differences subtraction A way to subtract in which differences are computed separately for each place (ones, tens, hundreds, and so on). The partial differences are then added to give the final answer.

$$\begin{array}{r} 932 \\ -\ 356 \\ \hline \end{array}$$

Subtract the hundreds:	$900 - 300 \rightarrow$	600
Subtract the tens:	$30 - 50 \rightarrow -$	20
Subtract the ones:	$2 - 6 \rightarrow -$	4
Find the total:	$600 - 20 - 4 \rightarrow$	576

partial-sums addition A way to add in which sums are computed for each place (ones, tens, hundreds, and so on) separately. The partial sums are then added to give the final answer.

$$\begin{array}{r} 496 \\ 229 \\ +\ 347 \\ \hline \end{array}$$

Add the hundreds:	$400 + 200 + 300 \rightarrow$	900
Add the tens:	$90 + 20 + 40 \rightarrow$	150
Add the ones:	$6 + 9 + 7 \rightarrow +$	22
Find the total:	$900 + 150 + 22 \rightarrow$	1,072

range The difference between the maximum and the minimum in a set of data.

trade-first subtraction A subtraction method in which all trades are done before any subtractions are carried out.

whole numbers The numbers 0, 1, 2, 3, 4, and so on.

Do-Anytime Activities

To work with your child on the concepts taught in this unit, try these interesting and rewarding activities:

1. Have your child see how many numbers he or she can identify in newspapers, magazines, advertisements, or news broadcasts.

2. Have your child collect and compare the measurements (height and weight) or accomplishments of favorite professional athletes.

3. Look up the different time zones of the United States and the world, quizzing your child on what time it would be at that moment at a particular location.

4. Have your child look for different representations of the same number. For example, he or she may see the same money amounts expressed in different ways, such as 50¢, $0.50, or 50 cents.

Building Skills through Games

In Unit 2, your child will play the following games. For detailed instructions, see the *Student Reference Book.*

Addition Top-It See *Student Reference Book,* page 263. This game provides practice with addition facts.

Fishing for Digits See *Student Reference Book,* page 242. This game provides practice identifying digits and the values of the digits, and adding and subtracting.

High-Number Toss See *Student Reference Book,* page 252. This game provides practice reading, writing, and comparing numbers.

Name That Number See *Student Reference Book,* page 254. This game reinforces skills in using all four operations.

Polygon Pair-Up See *Student Reference Book,* page 258. This game provides practice identifying properties of polygons.

Subtraction Target Practice See *Student Reference Book,* page 262. This game provides practice with subtraction and estimation.

Subtraction Top-It See *Student Reference Book,* pages 263 and 264. This is a variation of *Addition Top-It* and provides practice with subtraction facts.

As You Help Your Child with Homework

As your child brings assignments home, you may want to go over the instructions together, clarifying them as necessary. The answers listed below will guide you through this unit's Study Links.

Study Link 2·2

1. Sample answers: 8 × 8; 32 × 2; 10 + 54

2. Sample answers: 2 × 66; 11 × 12; 66 + 66; 30 + 30 + 30 + 30 + 12; (50 × 2) + 32

3. Sample answers: 20 + 20; 80 ÷ 2; $\frac{1}{2}$ × 80

4. Sample answers: 9 × 4; 72 ÷ 2; (12 × 4) − 12

Study Link 2·3

1. 876,504,000   2. 23,170,080

3. 876,504,000

4. a. thousand; 400,000

    b. million; 80,000,000

    c. million; 500,000,000

    d. thousand; 30,000

5. b. 596,708   d. 1,045,620

6. b. 13,877,000   d. 150,691,688

Study Link 2·4

2. 581,970,000   3. 97,654,320

5. a. 487,000,063   b. 15,000,297

6. 97,308,080

Study Link 2·5

2. 27   3. 8   4. 2   5. 6   6. 5

Study Link 2·6

1.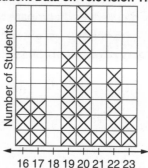

Student Data on Television Time

Number of Students (vertical axis)

16 17 18 19 20 21 22 23

Number of Hours Spent Watching Television Each Week

2. a. 23   b. 16   c. 7   d. 20   e. 20

4. 19.7

Study Link 2·7

1. 152   2. 510   3. 613

4. 1,432   5. 2,520   6. 5,747

11. 136   12. 720   13. 225

14. 720   15. 1,573   16. 2,356

Study Link 2·8

1. a. 645   b. 19   c. 626   d. 151

2. Giraffe, Asian elephant, and rhinoceros

3. 90   4. dog   5. mouse

Study Link 2·9

1. 68   11. 29

2. 382   12. 57

3. 367   13. 406

4. 3,746   14. 224

5. 2,889   15. 4,479

6. 2,322   16. 2,538

STUDY LINK 2·1 **Numbers Everywhere**

Find examples of numbers—all kinds of numbers. Look in newspapers and magazines. Look in books. Look on food packages. Ask people in your family for examples.

Write your numbers below. If an adult says you may, cut out the numbers and tape them onto the back of this page.

Be sure you write what the numbers mean.

Example: Mount Everest is 29,028 feet high. It is the world's tallest mountain.

_____

_____

_____

_____

_____

_____

_____

_____

_____

_____

_____

_____

_____

_____

_____

_____

_____

Practice

1. $5 \times 3 =$ _____ **2.** _____ $= 4 \times 3$ **3.** _____ $= 10 \div 2$ **4.** $8 \div 4 =$ _____

23

STUDY LINK 2·2   Many Names for Numbers

1. Write five names for 64.

64

2. Write five names for 132.

132

3. Pretend that the 4-key on your calculator is broken. Write six ways to display the number 40 on the calculator without using the 4-key. Try to use different numbers and operations.

     Example: $2 \times 2 \times 10$

_____ _____ _____

_____ _____ _____

Try This

4. Now pretend that all the keys on your calculator work except for the 3-key and the 6-key. Write six ways to display the number 36 without using these keys.

_____ _____ _____

_____ _____ _____

Practice

5. $20 + 60 =$ _____

6. _____ $= 60 + 90$

7. _____ $= 80 - 30$

8. $110 - 40 =$ _____

25

Place Value in Whole Numbers

1. Write the number that has

   6 in the millions place,
   4 in the thousands place,
   7 in the ten-millions place,
   5 in the hundred-thousands place,
   8 in the hundred-millions place, and
   0 in the remaining places.

   __ __ 6, __ __ __, __ __ __

2. Write the number that has

   7 in the ten-thousands place,
   3 in the millions place,
   1 in the hundred-thousands place,
   8 in the tens place,
   2 in the ten-millions place, and
   0 in the remaining places.

   __ __, __ __ __, __ __ __

3. Compare the two numbers you wrote in Problems 1 and 2.

   Which is greater? _____

4. The 6 in 46,711,304 stands for 6 __*million*__, or __6,000,000__.

   a. The 4 in 508,433,529 stands for 400 _____, or _____.

   b. The 8 in 182,945,777 stands for 80 _____, or _____.

   c. The 5 in 509,822,119 stands for 500 _____, or _____.

   d. The 3 in 450,037,111 stands for 30 _____, or _____.

Try This

5. Write the number that is 1 hundred thousand more.

   a. 210,366 __*310,366*__   b. 496,708 _____

   c. 321,589 _____   d. 945,620 _____

6. Write the number that is 1 million more.

   a. 3,499,702 __*4,499,702*__   b. 12,877,000 _____

   c. 29,457,300 _____   d. 149,691,688 _____

Practice

7. 32, 45, 58, _____, _____, _____

   Rule: _____

8. _____, _____, _____, 89, 115, 141

   Rule: _____

STUDY LINK 2·4 | **Place Values in Whole Numbers**

1. Write the numbers in order from smallest to largest.

15,964   1,509,460   150,094,400
1,400,960   15,094,600

_____

_____

_____

_____

_____

2. Write the number that has

5 in the hundred-millions place,
7 in the ten-thousands place,
1 in the millions place,
9 in the hundred-thousands place,
8 in the ten-millions place, and
0 in all other places.

— — —, — — —, — — —

3. Write the largest number you can. Use each digit just once.

3 5 0 7 9 2 6 4   _____

4. Write the value of the digit 8 in each numeral below.

a. 80,007,941 _____   **b.** 835,099,714 _____

c. 8,714,366 _____   **d.** 860,490 _____

5. Write each number using digits.

a. four hundred eighty-seven million, sixty-three _____

b. fifteen million, two hundred ninety-seven _____

Try This

6. I am an 8-digit number.
- The digit in the thousands place is the result of dividing 64 by 8.
- The digit in the millions place is the result of dividing 63 by 9.
- The digit in the ten-millions place is the result of dividing 54 by 6.
- The digit in the tens place is the result of dividing 40 by 5.
- The digit in the hundred-thousands place is the result of dividing 33 by 11.
- All the other digits are the result of subtracting any number from itself.

What number am I? __ __, __ __ __, __ __ __

STUDY LINK 2·5 | **Collecting Data**

1. Make a list of all the people in your family. Include all the people living at home now. Also include any brothers or sisters who live somewhere else. The people who live at home do not have to be related to you. Do not forget to write your name in the list.

   You will need this information to learn about the sizes of families in your class.

   _____ _____ _____

   _____ _____ _____

   _____ _____ _____

   How many people are in your family? _____ people

The tally chart at the right shows the number of books that some students read over the summer. Use the information to answer the questions below.

Number of Books Reported	Number of Students
2	///
3	//////
4	
5	/////////
6	////////
7	//
8	////

2. How many students reported the number of books they read? _____

3. What is the **maximum** (the largest number of books reported)? _____

4. What is the **minimum** (the smallest number of books reported)? _____

5. What is the **range?** _____

6. What is the **mode** (the most frequent number of books reported)? _____

Practice

7. 30 + 50 = _____

8. _____ = 70 + 70 + 70

9. _____ = 90 + 80 + 60

10. 100 + 40 + 70 = _____

31

Line Plots

The students in Sylvia's class estimated how much time they spend watching television each week. The tally chart below shows the data they collected.

Number of Hours per Week Spent Watching TV	Number of Students
16	///
17	///
18	
19	̶H̶H̶ /
20	̶H̶H̶ ////
21	/
22	̶H̶H̶
23	//

1. Construct a line plot for the data.

Student Data on Television Time

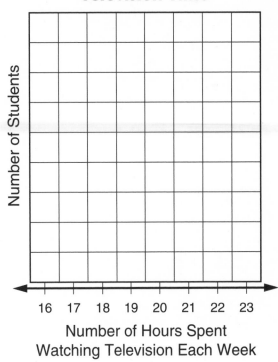

Number of Students

16 17 18 19 20 21 22 23

Number of Hours Spent
Watching Television Each Week

2. Find the following landmarks for the data:

   a. The maximum number of hours spent watching television each week. _____ hours

   b. minimum _____ hours   c. range _____ hours

   d. mode _____ hours   e. median _____ hours

3. Estimate the amount of time that you watch television each week. _____ hours

---

Try This

4. Calculate the mean number of hours Sylvia and her classmates spent

   watching TV each week. _____ hours

Practice

5. 80 + 30 = _____   6. _____ = 90 + 90

7. _____ = 70 + 60   8. 120 + 30 = _____

STUDY LINK 2·7 | **Multidigit Addition**

Make a ballpark estimate. Use the **partial-sums method** to add. Compare your answer with your estimate to see if your answer makes sense.

1.

$$67$$
$$+ 85$$

Ballpark estimate:

_____

2.

$$439$$
$$+ 71$$

Ballpark estimate:

_____

3.

$$227$$
$$+ 386$$

Ballpark estimate:

_____

4.

$$493$$
$$+ 939$$

Ballpark estimate:

_____

5.

$$732$$
$$+ 1,788$$

Ballpark estimate:

_____

6.

$$4,239$$
$$+ 1,508$$

Ballpark estimate:

_____

Practice

7. $8 \times 7 =$ _____

8. $9 \times 9 =$ _____

9. _____ $\div 6 = 9$

10. _____ $\div 4 = 8$

STUDY LINK 2·7

Multidigit Addition *continued*

Make a ballpark estimate. Use the **column-addition method** to add.
Compare your answer with your estimate to see if your answer makes sense.

SRB
11

11.	12.	13.
89 + 47	634 + 86	148 + 77
Ballpark estimate: _____	Ballpark estimate: _____	Ballpark estimate: _____
14.	15.	16.
481 + 239	746 + 827	508 + 1,848
Ballpark estimate: _____	Ballpark estimate: _____	Ballpark estimate: _____

Practice

17. 16, 21, 26, _____, _____, _____   Rule: _____

18. _____, 52, _____, 104, 130, _____   Rule: _____

STUDY LINK
2·8

Gestation Period

The period between the time an animal becomes pregnant and the time
its baby is born is called the **gestation period.** The table below shows the
number of days in the average gestation period for some animals.

1. For the gestation periods listed in the table ...

a. what is the maximum number of days?

_____ days

b. what is the minimum number of days?

_____ days

c. what is the range (the difference between
the maximum and the minimum)?

_____ days

d. what is the median (middle) number of days?

_____ days

Average Gestation Period (in days)	
Animal	**Number of Days**
dog	61
giraffe	457
goat	151
human	266
Asian elephant	645
mouse	19
squirrel	44
rhinoceros	480
rabbit	31

Source: World Almanac

2. Which animals have an average gestation period that is longer than 1 year?

_____

3. How much longer is the average gestation period for a goat than for a dog? _____ days

4. Which animal has an average gestation period that is about twice as long

as a rabbit's? _____

5. Which animal has an average gestation period that is about half as long

as a squirrel's? _____

| **Practice** |

6. 56 + 33 = _____

7. _____ = 167 + 96

8. _____ = 78 − 32

9. 271 − 89 = _____

37

Multidigit Subtraction

Make a ballpark estimate. Use the **trade-first subtraction method** to subtract.
Compare your answer with your estimate to see if your answer makes sense.

SRB
12

1.
 	
    96
 − 28
	2.
	
	    469
	 − 87
		3.
		
		    732
		 − 365

Ballpark estimate: _____	Ballpark estimate: . _____	Ballpark estimate: _____

4.
 	
  4,321
 − 575
	5.
	
	  5,613
	 − 2,724
		6.
		
		  6,600
		 − 4,278

Ballpark estimate: _____	Ballpark estimate: _____	Ballpark estimate: _____

Practice

7. 8 × _____ = 64 **8.** 9 × _____ = 72 **9.** 56 = _____ × 8 **10.** 42 = _____ × 7

STUDY LINK 2·9 Multidigit Subtraction *continued*

Make a ballpark estimate. Use the **partial-differences method** to subtract.
Compare your answer with your estimate to see if your answer makes sense.

11.	12.	13.
84 − 55	136 − 79	573 − 167
Ballpark estimate: _____	Ballpark estimate: _____	Ballpark estimate: _____
14.	**15.**	**16.**
506 − 282	5,673 − 1,194	3,601 − 1,063
Ballpark estimate: _____	Ballpark estimate: _____	Ballpark estimate: _____

Practice

17. _____, _____, 55, 44, _____, 22   Rule: _____

18. _____, _____, _____, _____, 72, 81   Rule: _____

STUDY LINK 2·10

Unit 3: Family Letter

Multiplication and Division; Number Sentences and Algebra

One of our goals in the coming weeks is to finish memorizing the multiplication facts for single-digit numbers. To help students master the facts, they will play several math games. Ask your child to teach you one of the games described in the *Student Reference Book,* and play a few rounds together.

The class will also take a series of 50-facts tests for multiplication. Because correct answers are counted only up to the first mistake (and not counted thereafter), your child may at first receive a low score. If this happens, don't be alarmed. Before long, scores will improve dramatically. Help your child set a realistic goal for the next test, and discuss what can be done to meet that goal.

Your child will use Multiplication/Division Fact Triangles to review the relationship between multiplication and division. (For example, 4 × 5 = 20, so 20 ÷ 5 = 4 and 20 ÷ 4 = 5.) You can use the triangles to quiz your child on the basic facts and test your child's progress.

In this unit, alternative symbols for multiplication and division are introduced. An asterisk (∗) may be substituted for the traditional × symbol, as in 4 ∗ 5 = 20. A slash (/) may be used in place of the traditional ÷ symbol, as in 20/4 = 5.

In Unit 3, the class will continue the World Tour, a yearlong project in which the students travel to a number of different countries. Their first flight will take them to Cairo, Egypt. These travels serve as background for many interesting activities in which students look up numerical information, analyze this information, and solve problems.

Finally, the class will have its first formal introduction to solving equations in algebra. (Informal activities with missing numbers in number stories have been built into the program since first grade.) Formal introduction to algebra in fourth grade may surprise you, because algebra is usually regarded as a high school subject. However, an early start in algebra is integral to the *Everyday Mathematics* philosophy.

Please keep this Family Letter for reference as your child works through Unit 3.

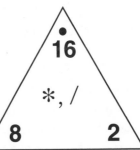

41

Vocabulary

Important terms in Unit 3:

dividend In division, the number that is being divided. For example, in $35 \div 5 = 7$, the dividend is 35.

divisor In division, the number that divides another number. For example, in $35 \div 5 = 7$, the divisor is 5.

Fact family A set of related arithmetic facts linking two inverse operations. For example, $4 + 8 = 12$, $8 + 4 = 12$, $12 - 4 = 8$, and $12 - 8 = 4$ is an addition/subtraction fact family, and $4 * 8 = 32$, $8 * 4 = 32$, $32/4 = 8$, and $32/8 = 4$ is a multiplication/ division fact family.

Fact Triangle A triangular flash card labeled with the numbers of a *fact family* that students can use to practice addition/subtraction or multiplication/ division facts.

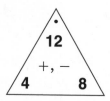 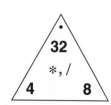

factor One of two or more numbers that are multiplied to give a product. For example, $4 * 1.5 = 6$; so 6 is the product, and 4 and 1.5 are the factors. See also *factor of a counting number* n.

factor of a counting number *n* A counting number whose product with some other counting number equals *n*. For example, 2 and 3 are factors of 6 because $2 * 3 = 6$. But 4 is not a factor of 6 because $4 * 1.5 = 6$ and 1.5 is not a counting number.

multiple of a number *n* A product of *n* and a counting number. The multiples of 7, for example, are 7, 14, 21, 28, and so on.

number sentence Two numbers or expressions separated by a relation symbol ($=, >, <, \geq, \leq,$ or $\neq$). Most number sentences also contain at least one operation symbol ($+, -, \times, *, \cdot, \div, /$). Number sentences may also have grouping symbols, such as parentheses.

open sentence A *number sentence* in which one or more *variables* hold the places of missing numbers. For example, $5 + x = 13$ is an open sentence.

percent (%) Per hundred, or out of a hundred. For example, "48% of the students in the school are boys" means that, on average, 48 out of every 100 students in the school are boys; $48\% = \frac{48}{100} = 0.48$

product The result of multiplying two numbers called *factors*. For example, in $4 * 3 = 12$, the product is 12.

quotient The result of dividing one number by another number. For example, in $35 \div 5 = 7$, the quotient is 7.

square number A number that is the product of a counting number and itself. For example, 25 is a square number because $25 = 5 * 5$. The square numbers are 1, 4, 9, 16, 25, and so on.

variable A letter or other symbol that represents a number. A variable can represent one specific number. For example, in the number sentence $5 + n = 9$, only *n* makes the sentence true. A variable may also stand for many different numbers. For example, $x + 2 < 10$ is true if *x* is any number less than 8. And in the equation $a + 3 = 3 + a$, *a* stands for all numbers.

"What's My Rule?" problem A type of problem that asks for a rule for relating two sets of numbers. Also, a type of problem that asks for one of the sets of numbers, given a rule and the other set of numbers.

Rule ×8	in	out
	6	48
	10	80
	3	
		56
		64

42

Do-Anytime Activities

To work with your child on the concepts taught in this unit, try these interesting and rewarding activities:

1. Continue to work on multiplication and division facts by using Fact Triangles and fact families and by playing games described in the *Student Reference Book.*

2. As the class proceeds through the unit, give your child multidigit addition and subtraction problems related to the lessons covered, such as 348 + 29, 427 + 234, 72 − 35, and 815 − 377.

3. Help your child recognize and identify real-world examples of right angles, such as the corner of a book, and examples of parallel lines, such as railroad tracks.

Building Skills through Games

In Unit 3, your child will play the following games.

Baseball Multiplication See *Student Reference Book,* pages 231 and 232.

Two players will need 4 regular dice, 4 pennies, and a calculator to play this game. Practicing the multiplication facts for 1–12 and strengthening mental arithmetic skills are the goals of *Baseball Multiplication.*

Beat the Calculator See *Student Reference Book,* page 233.

This game involves 3 players and requires a calculator and a deck of number cards, four each of the numbers 1 through 10. Playing *Beat the Calculator* helps your child review basic multiplication facts.

Division Arrays See *Student Reference Book,* page 240.

Materials for this game include number cards, 1 each of the numbers 6 through 18; a regular (6-sided) die; 18 counters; and paper and pencil. This game, involving 2 to 4 players, reinforces the idea of dividing objects into equal groups.

Multiplication Top-It See *Student Reference Book,* page 264.

The game can be played with 2 to 4 players and requires a deck of cards, four each of the numbers 1 through 10. This game helps your child review basic multiplication facts.

Name That Number See *Student Reference Book,* page 254.

Played with 2 or 3 players, this game requires a complete deck of number cards and paper and pencil. Your child tries to name a target number by adding, subtracting, multiplying, and dividing the numbers on as many of the cards as possible.

As You Help Your Child with Homework

As your child brings assignments home, you may want to go over the instructions together, clarifying them as necessary. The answers listed below will guide you through some of the Study Links in this unit.

Study Link 3·1

1. 60, 230, 110, 280, 370
2. 110, 80, 310, 240, 390
3. 34, 675, 54; +46   4. 9, 50, 420; ×7
5. 2, 400, 2,000   6. Answers vary.
7. 115   8. 612   9. 1,440

Study Link 3·2

2. 1, 2, 3, 4, 6, 9, 12, 18, 36   3. 1, 16; 2, 8; 4, 4
4. 56   5. Sample answer: 4, 8, 12, 16   6. 53
7. 388   8. 765

Study Link 3·3

1. 24   2. 54   3. 28   4. 16
5. 45   6. 18   7. 40   8. 25
9. 48   11. 1, 2, 3, 6, 9, 18

Study Link 3·4

1. 6   2. 8   3. 6   4. 3
6. 20; 5   7. 18; 6   8. 49; 7   9. 9; 2
10. 7; 5   11. 7; 4
12. Sample answer: 10, 15, 20, 25
13. 1, 2, 3, 4, 6, 8, 12, 24

Study Link 3·5

1. 5   2. 7   3. 72   4. 10
5. 32   15. 1,646   16. 5,033
17. 289   18. 1,288

Study Link 3·6

3. **a.** T
4. about 128,921 miles;
   $132,000 - 3,079 = 128,921$
5. **a.** 4
6. 1, 2, 3, 4, 6, 12
7. Sample answers: 16, 24, 32, 40

Study Link 3·7

	Cities	Measurement on Map (inches)	Real Distance (miles)
1.	Cape Town and Durban	4	800
2.	Durban and Pretoria	$1\frac{3}{4}$	350
3.	Cape Town and Johannesburg	4	800
4.	Johannesburg and Queenstown	2	400
5.	East London and Upington	$2\frac{1}{2}$	500
6.	____ and ____	Answers vary.	

Study Link 3·8

1. $659 - 457 = 202$; 202
2. $1,545 + 2,489 = 4,034$; 4034
3. $700 - 227 = 473$; 473
4. $1,552 - 1,018 = 534$; 534
5. $624 + 470 + 336 = 1,430$; 1,430   6. 9
7. 6, 12, 18, 24, 30, 36, 42, 48, 54, 60

Study Link 3·9

1. F   2. F   3. T   4. T
5. F   6. T   7. T   8. ?
11. **b.** $7 * 8 = 56$ 12. 36, 60, 84; +12
13. 54, 216, 324; +54

Study Link 3·10

1. 27   2. 33   3. 1   4. 24
5. 37   6. 8   7. $3 * (6 + 4) = 30$
8. $15 = (20/4) + 10$ 9. $7 + (7 * 3) = 4 * 7$
10. $9 * 6 = (20 + 7) * 2$
11. $72 \div 9 = (2 * 3) + (18 \div 9)$
12. $35 \div (42 \div 6) = (10 - 6) + 1$ 13. ?
14. ?   15. F   16. T   17. F   18. T

44

STUDY LINK 3·1   "What's My Rule?"

SRB 162–166

Complete the "What's My Rule?" tables and state the rules.

1. in ↓

Rule
Add 40

↓ out

in	out
20	
190	
70	
240	
330	

2. in ↓

Rule
−60

↓ out

in	out
	50
	20
	250
	180
	330

3. Rule: _____

in	out
131	177
	80
104	150
629	
	100

4. Rule: _____

in	out
70	490
	63
	350
20	140
60	

Try This

5. Rule: There are 20 nickels in $1.00.

dollars	nickels
3	60
	40
5	100
20	
100	

6. Create your own.

Rule: _____

in	out

Practice

7. _____ = 47 + 68   **8.** 359 + 253 = _____   **9.** 787 + 653 = _____

STUDY LINK 3·2

Multiplication Facts

1. Complete the Multiplication/Division Facts Table below.

*,/	1	2	3	4	5	6	7	8	9	10
1						6				
2										
3	3		9							
4		8								
5										
6										
7		14								
8										
9										
10										

2. List all the *factors* of 36. _____

3. List the *factor pairs* of 16. _____ and _____, _____ and _____, _____ and _____

4. Name the *product* of 8 and 7. _____

5. Name four *multiples* of 4. _____, _____, _____, _____

| **Practice** |

6. _____ = 91 − 38   **7.** _____ = 630 − 242   **8.** 1,462 − 697 = _____

STUDY LINK 3·3 Fact Triangles

Complete these Multiplication/Division Fact Triangles.

1.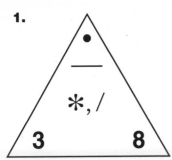

2.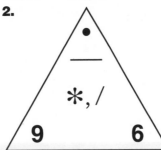

3.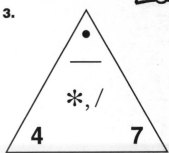

4.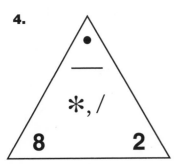

5.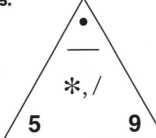

6.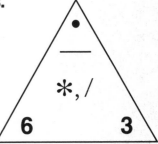

7.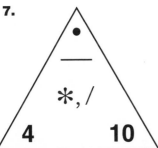

8.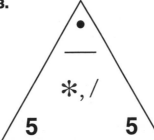

9.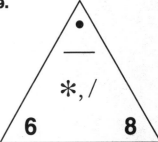

Practice

10. Name 4 multiples of 7. _____, _____, _____, _____

11. List all the factors of 18. _____

12. Name the product of 9 and 6. _____

13. List all the factor pairs of 20.

_____ and _____, _____ and _____, _____ and _____

49

Mystery Numbers

Find the mystery numbers.

1. I am thinking of a mystery number. If I multiply it by 4, the answer is 24. What is the number? _____

2. I am thinking of another number. If I multiply it by 3, the answer is 24. What is the number? _____

3. I multiplied a number by itself and got 36. What is the number? _____

4. If I multiply 7 by a number, I get 21. What is the number? _____

5. Write your own mystery number problem.

_____

_____

Fill in the missing numbers.

6. $4 * 5 =$ _____   _____ $* 4 = 20$

7. _____ $= 6 * 3$ $18 =$ _____ $* 3$

8. $7 * 7 =$ _____   _____ $* 7 = 49$

9. _____ $* 2 = 18$ $18 =$ _____ $* 9$

10. $35 =$ _____ $* 5$ _____ $* 7 = 35$

11. $28 =$ _____ $* 4$ _____ $* 7 = 28$

Practice

12. Name 4 multiples of 5. _____, _____, _____, _____

13. List all the factors of 24. _____

STUDY LINK 3·5   Missing Numbers

Complete each fact by filling in the missing numbers.
Use the Multiplication/Division Facts Table to help you.

1. 30 / 6 = _____

2. 21 / _____ = 3

3. 9 = _____ ÷ 8

4. 100 / _____ = 10

5. _____ / 4 = 8

6. 25 ÷ _____ = _____

7. _____ = 42 / _____

8. 8 / _____ = _____

9. 4 = _____ / _____

10. _____ ÷ _____ = 1

11. _____ / 2 = _____

12. 10 * _____ = _____

Try This

13. 5 * _____ * _____ = 30

14. 54 = _____ * _____ * _____

*,/	1	2	3	4	5	6	7	8	9	10
1	1	2	3	4	5	6	7	8	9	10
2	2	4	6	8	10	12	14	16	18	20
3	3	6	9	12	15	18	21	24	27	30
4	4	8	12	16	20	24	28	32	36	40
5	5	10	15	20	25	30	35	40	45	50
6	6	12	18	24	30	36	42	48	54	60
7	7	14	21	28	35	42	49	56	63	70
8	8	16	24	32	40	48	56	64	72	80
9	9	18	27	36	45	54	63	72	81	90
10	10	20	30	40	50	60	70	80	90	100

Practice

15. _____ = 989 + 657

16. 314 + 4,719 = _____

17. 887 − 598 = _____

18. _____ = 2,004 − 716

STUDY LINK
3·6

Number Stories about Egypt

1. The Nile in Africa is about 4,160 miles long. The Huang River in Asia is about 800 miles shorter than the Nile. How long is the Huang River?

   Number model: _____   About _____ miles

2. The Suez Canal links the Mediterranean and Red Seas. It is 103 miles long and was opened in 1869. For how many years has the Suez Canal been open?

   Number model: _____   _____ years

3. Egypt has about 3,079 miles of railroad. The United States has about 132,000 miles of railroad. How many fewer miles of railroad does Egypt have than the United States?

   Number model: _____   About _____ miles

4. The population of Cairo, the capital of Egypt, is about 10,834,000. The population of Washington, D.C., is about 563,000.

   a. True or false? About $10\frac{1}{2}$ million more people live in Cairo than in Washington, D.C.   _____

   b. Explain how you solved the problem.

   _____

   _____

Try This

5. The area of Egypt is about 386,700 square miles. The area of Wyoming is about 97,818 square miles.

   a. Egypt is about how many times as large as Wyoming? _____

   b. Explain how you solved the problem.

   _____

   _____

Practice

6. List all the factors of 12. _____

7. Name 4 multiples of 8. _____, _____, _____, _____

55

STUDY LINK 3·7 Map Scale

Here is a map of South Africa.
Use a ruler to measure the shortest
distance between cities. Measure
to the nearest $\frac{1}{4}$ inch. Use the map
scale to convert these measurements
to real distances.

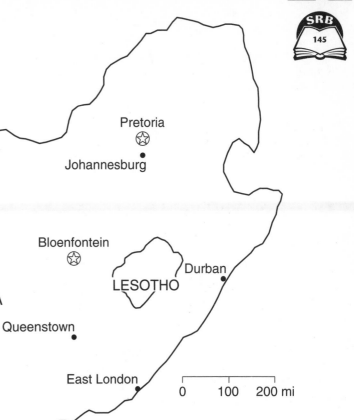

SRB
145

1 inch represents 200 miles

	Cities	Measurement on Map (inches)	Real Distance (miles)
1.	Cape Town and Durban		
2.	Durban and Pretoria		
3.	Cape Town and Johannesburg		
4.	Johannesburg and Queenstown		
5.	East London and Upington		
6.	_____ and _____		

Practice

7. _____ = 767 + 254   **8.** 193 + 6,978 = _____

9. 562 − 388 = _____   **10.** _____ = 4,273 − 678

57

STUDY LINK 3·8

Addition and Subtraction Number Stories

1. In 1896, the United Kingdom had the largest navy in the world with 659 ships. France had the second-largest navy with 457 ships. The United States was tenth with only 95 ships. How many more ships did the United Kingdom have than France?

_____ **Answer:** _____ more ships
                (number model)

2. Rhode Island, the smallest state in the United States, has an area of 1,545 square miles. The area of the second-smallest state, Delaware, is 2,489 square miles. What is the combined area of these two states?

_____ **Answer:** _____ square miles
                (number model)

3. A polar bear can weigh as much as 700 kilograms. An American black bear can weigh as much as 227 kilograms. How much more can a polar bear weigh than an American black bear?

_____ **Answer:** _____ kilograms more
                (number model)

4. The Pacific leatherback turtle's maximum weight is about 1,552 pounds. The Atlantic leatherback turtle's maximum weight is about 1,018 pounds. What is the difference between the turtles' weights?

_____ **Answer:** _____ pounds
                (number model)

5. According to the National Register of Historic Places, New York City has the most historic places in the United States with 624 sites. Philadelphia is second with 470 sites, and Washington, D.C., is third with 336 sites. How many historic sites are there in these three cities?

_____ **Answer:** _____ historic sites
                (number model)

Practice

6. The numbers 81, 27, and 45 are multiples of _____.

7. List the first ten multiples of 6.

_____, _____, _____, _____, _____, _____, _____, _____, _____, _____

STUDY LINK
3·9 **Number Sentences**

Next to each number sentence, write T if it is true, F if it is false, or ? if you can't tell.

1. $20 - 12 = 8 * 3$ _____

2. $7 = 14 * 2$ _____

3. $497 < 500$ _____

4. $16 / 4 = 4$ _____

5. $15 + 10 = 5$ _____

6. $24 > 11 + 11$ _____

7. $100 - 5 = 95$ _____

8. $33 - 4$ _____

9. Write two true number sentences. _____

_____

10. Write two false number sentences. _____

_____

11. **a.** Explain why $7 * 8$ is not a number sentence.

_____

_____

b. How could you change $7 * 8$ to make a true number sentence?

_____

_____

c. How could you change $7 * 8$ to make a false number sentence?

_____

_____

Practice

12. 24, _____, 48, _____, 72, _____   Rule: _____

13. _____, 108, 162, _____, 270, _____   Rule: _____

61

STUDY LINK 3·10 Parentheses in Number Sentences

Write the missing number to make each number sentence true.

1. (45 / 5) * 3 = _____

2. 9 + (4 * 6) = _____

3. (20 ÷ 4) ÷ 5 = _____

4. _____ = (33 − 25) * 3

5. _____ = (25 / 5) + (8 * 4)

6. (33 + 7) / (3 + 2) = _____

Insert parentheses ( ) to make each number sentence true.

7. 3 * 6 + 4 = 30

8. 15 = 20 / 4 + 10

9. 7 + 7 * 3 = 4 * 7

10. 9 * 6 = 20 + 7 * 2

Try This

Insert two sets of parentheses to make each number sentence true.

11. 72 ÷ 9 = 2 * 3 + 18 ÷ 9

12. 35 ÷ 42 ÷ 6 = 10 − 6 + 1

Write T if it is true, F if it is false, or ? if you can't tell.

13. (6 * 5) / 3 _____

14. (3 * 7) / (15 − 12) _____

15. 30 = 1 + (4 * 6) _____

16. (4 * 6) + 13 = 47 − 10 _____

17. 15 > (7 * 6) * (10 − 9) _____

18. 20 < (64 ÷ 8) * (12 ÷ 4) _____

Practice

19. _____ = 494 + 3,769

20. 5,853 + 4,268 = _____

21. _____ = 8,210 − 654

22. 7,235 − 906 = _____

STUDY LINK 3·11 | **Open Sentences**

Write T if the number sentence is true and F if the number sentence is false.

1. $35 = 7 * 5$ _____

2. $43 > 34$ _____

3. $25 + 25 < 50$ _____

4. $49 - (7 \times 7) = 0$ _____

Make a true number sentence by filling in the missing number.

5. _____ $= 12 / (3 + 3)$

6. $(60 - 28) / 4 =$ _____

7. $(3 \times 8) \div 6 =$ _____

8. $30 - (4 + 6) =$ _____

Make a true number sentence by inserting parentheses.

9. $4 * 2 + 10 = 18$

10. $16 = 16 - 8 * 2$

11. $27 / 9 / 3 = 1$

12. $27 / 9 / 3 = 9$

Find the solution of each open sentence below. Write a number sentence with the solution in place of the variable. Check to see whether the number sentence is true.

Example: $6 + x = 14$ **Solution:** 8 **Number sentence:** $6 + 8 = 14$

Open sentence	Solution	Number sentence
13. $12 + x = 32$	_____	_____
14. $s = 200 - 3$	_____	_____
15. $5 * y = 40$	_____	_____
16. $7 = x / 4$	_____	_____

Practice

17. $366 + 7,565 =$ _____

18. $3,238 + 9,784 =$ _____

19. $9,325 - 756 =$ _____

20. $4,805 - 2,927 =$ _____

**STUDY LINK
3·12** | **Unit 4: Family Letter**

Decimals and Their Uses

In previous grades, your child had many experiences with money written in decimal notation. In the next unit, the class will learn about other uses of decimals.

The class will focus on examples of decimals in everyday life. For example, some thermometers have marks that are spaced $\frac{2}{10}$ of a degree apart. These marks give a fairly precise measurement of body temperature, such as 98.6 °F.

Normal body temperature is about 98.6 °F.

Students will explore how decimals are used in measuring distances, times, and gasoline mileage.

We will also begin a yearlong measurement routine. Students will find their own "personal references," which they will use to estimate lengths, heights, and distances in metric units. For example, your child might discover that the distance from the base of his or her thumb to the tip of his or her index finger is about 10 centimeters and then use this fact to estimate other distances.

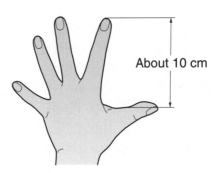

About 10 cm

The World Tour will continue. In small groups, students will gather information about different countries in Africa and then share what they have learned with the class. Students can then compare and interpret data for a large number of countries from the same region.

Please keep this Family Letter for reference as your child works through Unit 4.

Vocabulary

Important terms in Unit 4:

centimeter (cm) In the metric system, a unit of length equivalent to $\frac{1}{100}$ of a meter; 10 millimeters; $\frac{1}{10}$ of a decimeter.

decimeter (dm) In the metric system, a unit of length equivalent to $\frac{1}{10}$ of a meter; 10 centimeters.

hundredths In base-10 *place-value* notation, the place in which a digit has a value equal to $\frac{1}{100}$ of itself; the second digit to the right of the decimal point.

meter (m) In the metric system, the unit of length from which other units of length are derived. One meter is the distance light will travel in a vacuum (empty space) in $\frac{1}{299,792,458}$ second; 100 centimeters; 10 decimeters.

millimeter (mm) A metric unit of length equivalent to $\frac{1}{1,000}$ of a meter; $\frac{1}{10}$ of a centimeter.

ONE Same as *whole*.

ones The place-value position in which a digit has a value equal to the digit itself.

personal measurement reference A convenient approximation for a standard unit of measurement. For example, many people have thumbs that are approximately one inch wide.

place value A number writing system that gives a digit a value according to its position, or place, in the number. In our standard, base-10 system, each place has a value ten times that of the place to its right and 1 tenth the value of the place to its left.

1,000s	100s	10s	1s		0.1s	0.01s	0.001s
Thousands	Hundreds	Tens	Ones	.	Tenths	Hundredths	Thousandths

tens The place-value position in which a digit has a value equal to 10 times itself.

tenths In base-10 *place-value* notation, the place in which a digit has a value equal to $\frac{1}{10}$ of itself; the first digit to the right of the decimal point.

thousandths In base-10 *place-value* notation, the place in which a digit has a value equal to $\frac{1}{1,000}$ of itself; the third digit to the right of the decimal point.

whole (or ONE, or unit) In *Everyday Mathematics*, an entire object, collection of objects, or quantity being considered; 100%. Same as the ONE or unit whole.

Do-Anytime Activities

To work with your child on the concepts taught in this unit, try the interesting activities listed below. For each activity, discuss the use of decimals and the meanings of place values.

1. Have your child track the sports statistics of a favorite athlete.

2. Have your child compare prices of items in the supermarket.

3. Help your child create and use new personal reference measures.

4. Together, find statistics about countries in the World Tour. Look in newspapers and almanacs.

Building Skills through Games

In Unit 4, your child will play the following games.

Baseball Multiplication See *Student Reference Book*, pages 231 and 232. The game provides practice with multiplication facts.

Fishing for Digits See *Student Reference Book*, page 242. The game provides practice in identifying digits, the values of the digits, adding, and subtracting.

Name That Number See *Student Reference Book*, page 254. The game provides practice with using operations to represent numbers in different ways.

Number Top-It **(Decimals)** See *Student Reference Book*, page 256. The game provides practice with comparing, ordering, reading, and identifying the value of digits in decimal numbers.

Polygon Pair-Up See *Student Reference Book*, page 258. The game provides practice in identifying properties of polygons.

Product Pile-Up See *Student Reference Book*, page 259. The game provides practice with multiplication facts.

As You Help Your Child with Homework

As your child brings assignments home, you may want to go over the instructions together, clarifying them as necessary. The answers listed below will guide you through some of the Study Links in this unit.

Study Link 4·1

1.

1,000s	100s	10s	1s
6	8	5	4

3.

10s	1s		0.1s	0.01s	0.001s
7	3	.	0	0	4

Study Link 4·3

Sample answers:

3. 5.05, 5.25, 5.95

4. 4.15, 4.55, 4.99

5. 21.4, 21.98, 21.57

6. 0.89, 0.85, 0.82

7. 2.155, 2.16, 2.159

8. 0.84, 0.88, 0.87

Study Link 4·4

1. Seikan and Channel Tunnel

2. Between 90 and 130 miles

3. Sample answer: I rounded the tunnel lengths to "close-but-easier" numbers and added $35 + 30 + 20 + 15 + 15 = 115$ to find the total length.

4. 12 miles   5. 8 miles

Study Link 4·5

1. 120.41 2. 1.46   3. 5.18   4. 0.03

5. >   6. <   7. >   8. >

9. Sample answer: 2.33 + 4.21

10. Sample answer: 6.83 − 5.31

Study Link 4·6

1. a. $0.76 b. $2.43 c. $4.64 d. $2.95

2. $16.40   3. $2.57   4. $7.32   5. $18.10

6. $10.78

7. Loaf of bread; Sample answer: The price of a loaf of bread in 2000 was $0.88. The expected price of a loaf of bread in 2025 is $3.31. This was almost 4 times its cost in 2000.

Study Link 4·7

1. $\frac{335}{1,000}$; 0.335   2. $\frac{301}{1,000}$; 0.301

3. $\frac{7}{100}$; 0.07   4. $1\frac{5}{100}$; 1.05

5. 0.346 6. 0.092 7. 0.003 8. 2.7

9. 0.536 10. 0.23 11. 7.008 12. 0.4

13. >   14. >   15. <   16. <

Study Link 4·8

1. a. 7 cm b. 0.07 m 2. a. 12 cm b. 0.12 m

3. a. 4 cm b. 0.04 m 4. a. 6 cm b. 0.06 m

5. a. 2 cm b. 0.02 m 6. a. 14 cm b. 0.14 m

Study Link 4·9

2. 180 mm 3. 4 cm   4. 3,000 mm

5. 400 cm 6. 7 m   7. 460 cm

8. 794 cm 9. 4.5 m   10. 0.23 m

11. 60 cm 12. 8 cm   13. 7 cm

Study Link 4·10

2. a. 65 mm   b. 2.6 cm   c. 610 cm

3. a. 50 mm   b. 3 cm   c. 300 cm

4. a. 800 mm   b. 11 cm   c. 5 m

5. a. 430 mm   b. 9.8 cm   c. 0.34 m

6. a. 6 mm   b. 0.4 cm   c. 5,200 mm

STUDY LINK 4·1 | Place-Value Puzzles

Use the clues to write the digits in the boxes and find each number.

1. ◆ Write 5 in the tens place.

◆ Find $\frac{1}{2}$ of 24. Subtract 4. Write the result in the hundreds place.

◆ Add 7 to the digit in the tens place. Divide by 2. Write the result in the thousands place.

◆ In the ones place, write an even number greater than 2 that has not been used yet.

1,000s	100s	10s	1s

2. ◆ Divide 15 by 3. Write the result in the hundredths place.

◆ Multiply 2 by 10. Divide by 10.
Write the result in the ones place.

◆ Write a digit in the tenths place that is 4 more than the digit in the hundredths place.

◆ Add 7 to the digit in the ones place.
Write the result in the thousandths place.

100s	10s	1s	0.1s	0.01s	0.001s
		.			

3. ◆ Write the result of 6 ∗ 9 divided by 18 in the ones place.

◆ Double 8. Divide by 4. Write the result in the thousandths place.

◆ Add 3 to the digit in the thousandths place. Write the result in the tens place.

◆ Write the same digit in the tenths and hundredths place so that the sum of all the digits is 14.

10s	1s	0.1s	0.01s	0.001s
	.			

Practice

Write true or false.

4. 6 ∗ 5 = 15 + 15 _____

5. 15 + 7 < 13 − 8 _____

6. 72 / 9 > 9 _____

STUDY LINK 4·2

Decimals All Around

Find examples of decimals in newspapers, in magazines, in books, or on food packages. Ask people in your family for examples.

Write your numbers below or, if an adult says you may, cut them out and tape them on this page. Be sure to write what the numbers mean. For example, "The body temperature of a hibernating dormouse may go down to 35.6°F."

_____

_____

_____

_____

_____

_____

_____

_____

_____

_____

_____

_____

Practice

Write true or false.

1. $286 + 286 = 462$ _____

2. $907 - 709 = 200$ _____

3. $641 + 359 = 359 + 641$ _____

4. $2,345 - 198 = 2,969 - 822$ _____

Ordering Decimals

SRB
33

Mark the approximate locations of the decimals and fractions on the
number lines below. Rename fractions as decimals as necessary.

1.

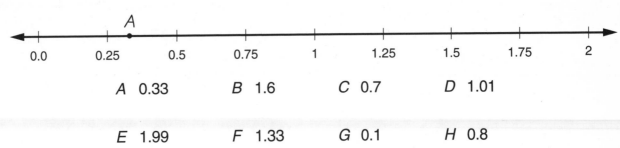

0.0	0.25	0.5	0.75	1	1.25	1.5	1.75	2

A   0.33   B   1.6   C   0.7   D   1.01

E   1.99   F   1.33   G   0.1   H   0.8

2.

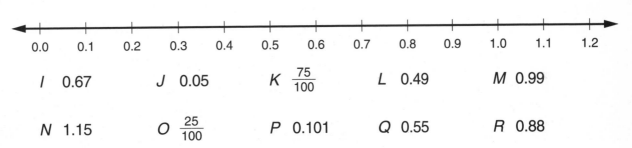

0.0	0.1	0.2	0.3	0.4	0.5	0.6	0.7	0.8	0.9	1.0	1.1	1.2

I   0.67   J   0.05   K $\frac{75}{100}$ L   0.49   M   0.99

N   1.15   O $\frac{25}{100}$ P   0.101   Q   0.55   R   0.88

Use decimals. Write 3 numbers that are between the following:

3. $5 and $6 $_____ $_____ $_____

4. 4 centimeters and
5 centimeters   _____ cm   _____ cm   _____ cm

5. 21 seconds and
22 seconds   _____ sec   _____ sec   _____ sec

6. 8 dimes and 9 dimes $_____ $_____ $_____

7. 2.15 meters and
2.17 meters   _____ m   _____ m   _____ m

8. 0.8 meter and 0.9 meter   _____ m   _____ m   _____ m

Practice

9. $x + 17 = 23$ $x =$ _____   **10.** $5 * n = 35$ $n =$ _____   **11.** $32 / b = 4$ $b =$ _____

STUDY LINK 4·4 | **Railroad Tunnel Lengths**

The table below shows the five longest railroad tunnels in the world.

Tunnel	Location	Year Completed	Length in Miles
Seikan	Japan	1988	33.46
Channel	France/England	1994	31.35
Moscow Metro	Russia	1979	19.07
London Underground	United Kingdom	1939	17.30
Dai-Shimizu	Japan	1982	13.98

Use estimation to answer the following questions.

1. Which two tunnels have a combined length of about 60 miles?

_____ and _____

2. Which of the following is closest to the combined length of all five tunnels? Choose the best answer.

⬭ Less than 90 miles   ⬭ Between 90 and 130 miles

⬭ Between 130 and 160 miles ⬭ More than 160 miles

3. Explain how you solved Problem 2.

_____

_____

4. About how many miles longer is the Channel Tunnel than the Moscow Metro Tunnel?

About _____ miles

Try This

5. The Cascade Tunnel in Washington State is the longest railroad tunnel in the United States. It is about $\frac{1}{4}$ the length of the Seikan. About how long is the Cascade Tunnel?

About _____ miles

Practice

6. $190 + b = 200$ $b =$ _____

7. $g - 500 = 225$ $g =$ _____

STUDY LINK 4·5 Addition and Subtraction of Decimals

Add or subtract. Show your work.

1. 96.45 + 23.96 = _____

2. 1.06 + 0.4 = _____

3. 9.87 − 4.69 = _____

4. 0.4 − 0.37 = _____

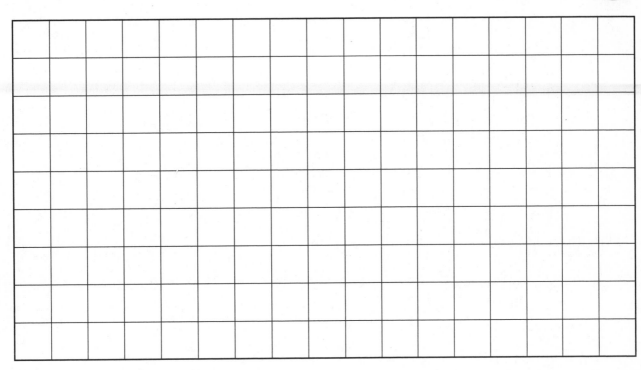

Write <, >, or = to make each statement true.

5. 2.78 + 9.1 _____ 3.36 + 8.49

6. 0.08 + 0.97 _____ 1.04 + 0.03

7. 13.62 − 4.9 _____ 9.4 − 1.33

8. 9.4 − 5.6 _____ 8.3 − 4.7

9. Name two 3-digit numbers whose sum is 6.54. _____ + _____ = 6.54

10. Name two 3-digit numbers whose difference is 1.52. _____ − _____ = 1.52

Practice

11. 13 = 7 + s   s = _____

12. 8 * g = 24   g = _____

13. 36 / p = 6   p = _____

14. m / 9 = 8   m = _____

STUDY LINK 4·6 | **Rising Grocery Prices**

The table below shows some USDA grocery prices for the year 2000 and estimates of grocery prices for the year 2025.

Grocery Item	Price in 2000	Estimated Price in 2025
dozen eggs	$1.02	$1.78
loaf of white bread	$0.88	$3.31
pound of butter	$2.72	$7.36
gallon of milk	$2.70	$5.65

1. How much more is each item predicted to cost in 2025?

   a. eggs _____ **b.** bread _____ **c.** butter _____ **d.** milk _____

2. The year is 2000. You buy bread and butter. You hand the cashier a $20 bill. How much change should you receive? _____

3. The year is 2025. You buy eggs and milk. You hand the cashier a $10 bill. How much change should you receive? _____

4. The year is 2000. You buy all 4 items. What is the total cost? _____

5. The year is 2025. You buy all 4 items. What is the total cost? _____

6. If the predictions are correct, how much more will you pay in 2025 for the 4 items than you paid in 2000? _____

7. Which item is expected to have the greatest price increase? _____

   Explain your answer. _____

   _____

   _____

Practice

8. List the first ten multiples of 3. ____, ____, ____, ____, ____, ____, ____, ____, ____, ____

9. List the first ten multiples of 7. ____, ____, ____, ____, ____, ____, ____, ____, ____, ____

81

STUDY LINK
4·7 Tenths, Hundredths, Thousandths

Complete the table. The big cube is the ONE.

Base-10 Blocks	Fraction Notation	Decimal Notation
1. □□□ ‖‖‖		
2. □□□ .		
3. ‖‖‖‖‖ ‖		
4. ◻⃞ ‖‖‖‖		

Write each number in decimal notation.

5. $\frac{346}{1,000}$ _____   **6.** $\frac{92}{1,000}$ _____

7. $\frac{3}{1,000}$ _____   **8.** $2\frac{7}{10}$ _____

Write each of the following in decimal notation.

9. 536 thousandths _____   **10.** 23 hundredths _____

11. 7 and 8 thousandths _____   **12.** 4 tenths _____

Write < or >.

13. 0.407 _____ 0.074   **14.** 0.65 _____ 0.437

15. 0.672 _____ 0.7   **16.** 2.38 _____ 2.4

Practice

17. 6.05 + 1.24 = _____   **18.** _____ = 47.90 + 0.76

19. _____ = 8.71 − 2.78   **20.** 46.8 − 3.77 = _____

STUDY LINK
4·8 | **Measuring in Centimeters**

Measure each line segment to the nearest centimeter. Record the measurement in centimeters and meters.

SRB
128 129

Example: _____

    a. About ____5____ centimeters   **b.** About __0.05__ meter

1. _____

    a. About _____ centimeters   **b.** About _____ meter

2. _____

    a. About _____ centimeters   **b.** About _____ meter

3. _____

    a. About _____ centimeters   **b.** About _____ meter

4. _____

    a. About _____ centimeters   **b.** About _____ meter

5. _____

    a. About _____ centimeters   **b.** About _____ meter

6. _____

    a. About _____ centimeters   **b.** About _____ meter

Practice

7. _____ = 10.06 + 10.04   **8.** 38.93 + 92.4 = _____

9. 16.85 − 14.23 = _____   **10.** _____ = 20.9 − 8.57

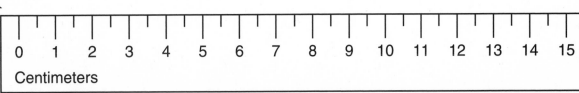

85

STUDY LINK 4·9

Metric Measurements

SRB
129 130

1. Use your personal references to estimate the lengths of 4 objects in metric units. Then measure each object. Record your estimates and measurements.

Object	Estimated Length	Actual Length

Complete.

2. 18 cm = _____ mm

3. _____ cm = 40 mm

4. 3 m = _____ mm

5. 4 m = _____ cm

6. _____ m = 700 cm

7. 4.6 m = _____ cm

8. 7.94 m = _____ cm

9. _____ m = 450 cm

10. _____ m = 23 cm

11. 0.6 m = _____ cm

Measure each line segment to the nearest $\frac{1}{2}$ cm.

12. _____

About _____ centimeters

13. _____

About _____ centimeters

Practice

Insert < or >.

14. 0.68 _____ 0.32

15. 9.13 _____ 9.03

16. 0.65 _____ 0.6

87

Decimals and Metric Units

Symbols for Metric Units of Length
meter (m)
centimeter (cm)
decimeter (dm)
millimeter (mm)

1 decimeter

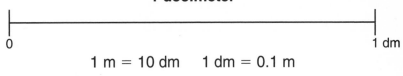

0   1 dm

1 m = 10 dm   1 dm = 0.1 m

10 centimeters

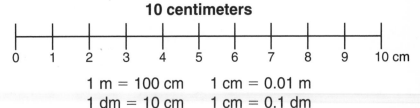

0   1   2   3   4   5   6   7   8   9   10 cm

1 m = 100 cm   1 cm = 0.01 m
1 dm = 10 cm   1 cm = 0.1 dm

100 millimeters

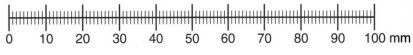

0   10   20   30   40   50   60   70   80   90   100 mm

1 m = 1,000 mm   1 mm = 0.001 m
1 dm = 100 mm   1 mm = 0.01 dm
1 cm = 10 mm   1 mm = 0.1 cm

Use your tape measure or ruler to help you fill in the answers below.

1. a. 4.2 cm = __42__ mm   **b.** 64 mm = __6.4__ cm   **c.** 2.6 m = __260__ cm

2. a. 6.5 cm = _____ mm   **b.** 26 mm = _____ cm   **c.** 6.1 m = _____ cm

3. a. 5 cm = _____ mm   **b.** 30 mm = _____ cm   **c.** 3 m = _____ cm

4. a. 80 cm = _____ mm   **b.** 110 mm = _____ cm   **c.** _____ m = 500 cm

5. a. 43 cm = _____ mm   **b.** 98 mm = _____ cm   **c.** _____ m = 34 cm

6. a. 0.6 cm = _____ mm   **b.** 4 mm = _____ cm   **c.** 5.2 m = _____ mm

---

Practice

7. 21, 49, and 56 are multiples of _____.

8. 45, 63, and 18 are multiples of _____.

89

STUDY LINK 4·11 | Unit 5: Family Letter

Big Numbers, Estimation, and Computation

In this unit, your child will begin to multiply 1- and 2-digit numbers using what we call the **partial-products method.** In preparation for this, students will learn to play the game *Multiplication Wrestling*. Ask your child to explain the rules to you and play an occasional game together. While students are expected to learn the partial-products method, they will also investigate the **lattice multiplication method,** which students have often enjoyed in the past.

If your child is having trouble with multiplication facts, give short (five-minute) reviews at home, concentrating on the facts he or she finds difficult.

Another important focus in this unit is on reading and writing big numbers. Students will use big numbers to solve problems and make reasonable estimates. Help your child locate big numbers in newspapers and other sources, and ask your child to read them to you. Or, you can read the numbers and have your child write them.

Sometimes it is helpful to write big numbers in an abbreviated form so that they are easier to work with. One way is to use **exponents,** which tell how many times a number, called the base, is used as a factor. For example, 100,000 is equal to $10 * 10 * 10 * 10 * 10$. So 100,000 can be written as 10^5. The small raised 5 is called an exponent, and 10^5 is read as "10 to the fifth power." This will be most students' first experience with exponents, which will be studied in depth during fifth and sixth grades.

The class is well into the World Tour. Students are beginning to see how numerical information about a country helps them get a better understanding of the country—its size, climate, location, and population distribution—and how these characteristics affect the way people live. The next stop on the World Tour will be Budapest, Hungary, the starting point for an exploration of European countries. Encourage your child to bring to school materials about Europe, such as articles in the travel section of your newspaper, magazine articles, and travel brochures.

Please keep this Family Letter for reference as your child works through Unit 5.

Vocabulary

Important terms in Unit 5:

billion 1,000,000,000, or 10^9; 1,000 million.

estimate A close, rather than exact, answer; an approximate answer to a computation; a number close to another number.

exponent See *exponential notation.*

exponential notation A way to show repeated multiplication by the same factor. For example, 2^3 is exponential notation for $2 * 2 * 2$. The small, raised 3 is the exponent. It tells how many times the number 2, called the base, is used as a factor.

$2^3 \leftarrow$ exponent
$\llcorner$ base

extended multiplication fact A multiplication fact involving multiples of 10, 100, and so on. In an extended multiplication fact, each factor has only one digit that is not 0. For example, $400 * 6 = 2,400$ and $20 * 30 = 600$ are extended multiplication facts.

lattice multiplication A very old way to multiply multidigit numbers. The steps below show how to find the product $46 * 73$ using lattice multiplication.

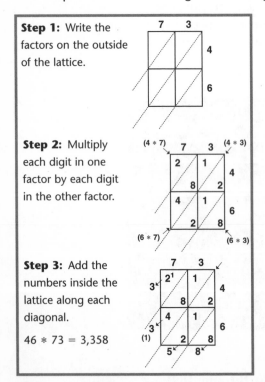

Step 1: Write the factors on the outside of the lattice.

Step 2: Multiply each digit in one factor by each digit in the other factor.

Step 3: Add the numbers inside the lattice along each diagonal.

$46 * 73 = 3,358$

magnitude estimate A rough estimate of whether a number is in the 1s, 10s, 100s, 1,000s, and so on.

million 1,000,000, or 10^6; 1,000 thousand.

partial-products multiplication A way to multiply in which the value of each digit in one factor is multiplied by the value of each digit in the other factor. The final product is the sum of the partial products. The example shows how to use the method to find $73 * 46$.

> **Partial–Products Multiplication**
> Multiply each part of one factor by each part of the other factor. Then add the partial products.
>
> $$\begin{array}{rr} & 73 \\ * & 46 \\ \hline 40 * 70 \rightarrow & 2,800 \\ 40 * 3 \ \rightarrow & 120 \\ 6 * 70 \rightarrow & 420 \\ 6 * 3 \ \rightarrow + & 18 \\ \hline & 3,358 \end{array}$$

power of 10 A whole number that can be written as a product using only 10s as factors. For example, 100 is equal to $10 * 10$, or 10^2. 100 is 10 to the second power or the second power of 10 or 10 squared.

round a number To approximate a number to make it easier to work with or to make it better reflect the precision of data. Often, numbers are rounded to a nearest *power of 10*. For example, 12,964 rounded to the nearest thousand is 13,000.

Do-Anytime Activities

To work with your child on concepts taught in this unit, try these interesting and rewarding activities:

1. To help your child practice handling big numbers, have him or her look up the distances from Earth to some of the planets in the solar system, such as the distance from Earth to Mars, to Jupiter, to Saturn, and so on.

2. Have your child look up the box-office gross of one or more favorite movies.

3. Help your child look up the populations and land areas of the state and city in which you live and compare them with the populations and areas of other states and cities.

4. Have your child locate big numbers in newspapers and other sources and ask him or her to read them to you. Or, you can read the numbers and have your child write them.

Building Skills through Games

In Unit 5, your child will practice multiplication skills and build his or her understanding of multidigit numbers by playing the following games. For detailed instructions, see the *Student Reference Book.*

Beat the Calculator See *Student Reference Book* page 233.
This game develops automaticity with extended multiplication facts.

High-Number Toss See *Student Reference Book* page 252.
This game reinforces understanding of place value.

Multiplication Wrestling See *Student Reference Book* page 253.
This game reinforces understanding of the partial-products method for multiplication.

Number Top-It See *Student Reference Book* page 255.
This game strengthens understanding of place value.

Product Pile Up See *Student Reference Book* page 259.
This game develops automaticity with multiplication facts.

As You Help Your Child with Homework

As your child brings assignments home, you may want to go over the instructions together, clarifying them as necessary. The answers listed below will guide you through some of the Study Links in this unit.

Study Link 5·1

9. 1.48 **10.** 1.13 **11.** 8.17

Study Link 5·2

1. 42; 420; 420; 4,200; 4,200; 42,000

2. 27; 270; 270; 2,700; 2,700; 27,000

3. 32; 320; 320; 3,200; 3,200; 32,000

4. 3; 5; 50; 3; 3; 500

5. 6; 6; 60; 9; 900; 9,000

6. 5; 500; 50; 8; 80; 800

7. 15 **8.** 9.5 **9.** 4.26

Study Link 5·3

Sample answers:

1. 850 + 750 = 1,600; 1,601

2. 400 + 1,000 + 500 = 1,900; 1,824

3. 400 + 750 = 1,150

4. 600 + 650 + 350 = 1,600; 1,595

5. 300 + 300 + 500 = 1,100

6. 800 + 700 = 1,500; 1,547

7. 700 + 200 + 400 = 1,300

8. 100 + 700 + 800 = 1,600; 1,627

9. 750 + 400 + 200 = 1,350

10. 600 + 800 = 1,400

11. 4,800 **12.** 2,100 **13.** 45,000

Study Link 5·4

Sample answers:

1. 20 * 400 = 8,000; 1,000s

2. 10 * 20 = 200; 100s

3. 5 * 400 = 2,000; 1,000s

4. 2 * 20 * 10,000 = 400,000; 100,000s

5. Either 3 or 4 digits; 10 * 10 = 100 and 90 * 90 = 8,100

Study Link 5·5

1. 392 **2.** 2,200 **3.** 11,916

4. a. 7 * 200 = 1,400; 1,000s **b.** 1,267 hours

5. less **6.** 7,884 **7.** 11,436

8. 1,258 **9.** 4,689

Study Link 5·6

1. 4,074 **2.** 1,680 **3.** 2,100 **4.** 486

5. 3,266 **6.** 17,000 **7.** 7,471 **8.** 37,632

9. 5,722 **10.** 10,751 **11.** 916 **12.** 2,769

Study Link 5·7

7. 6,552

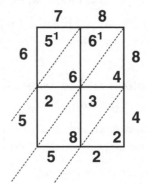

9. 39.57 **10.** 74.22 **11.** 33.77 **12.** 71.15

Study Link 5·8

92,106,954,873

12. 92 billion, 106 million, 954 thousand, 873

13. 370 **14.** 3,168 **15.** 1,656 **16.** 2,632

Study Link 5·9

7. 441 **8.** 2,970 **9.** 5,141

Study Link 5·10

2. Phoenix Mercury and San Antonio Stars; Sacramento Monarchs and Seattle Storm

4. 4,152 **5.** 798 **6.** 3,212

Study Link 5·11

1. China **2.** France **4.** Italy and the United States

STUDY LINK
5·1

Multiplication/Division Puzzles

Solve the multiplication/division puzzles mentally. Fill in the blank boxes.

Examples:

*, /	300	2,000
2	600	4,000
3	900	6,000

*, /	80	50
4	320	200
8	640	400

1.

*, /	70	400
8		
9		

2.

*, /	5	7
80		
600		

3.

*, /	9	4
50		
7,000		

4.

*, /		600
7	3,500	
		2,400

5.

*, /		80
30	2,700	
		56,000

6.

*, /	4,000	
	36,000	
20		10,000

Make up and solve some puzzles of your own.

7.

*, /		

8.

*, /		

Practice

9. _____ = 0.56 + 0.92

10. _____ = 2.86 − 1.73

11. 19.11 − 10.94 = _____

12. _____ = 0.52 + 0.25

95

STUDY LINK 5·2 | Extended Multiplication Facts

Solve mentally.

1. 6 * 7 = _____

6 * 70 = _____

60 * 7 = _____

60 * 70 = _____

600 * 7 = _____

60 * 700 = _____

2. 9 * 3 = _____

9 * 30 = _____

90 * 3 = _____

90 * 30 = _____

900 * 3 = _____

90 * 300 = _____

3. 4 * 8 = _____

4 * 80 = _____

40 * 8 = _____

40 * 80 = _____

400 * 8 = _____

40 * 800 = _____

4. 5 * _____ = 15

30 * _____ = 150

30 * _____ = 1,500

_____ * 50 = 150

_____ * 500 = 1,500

30 * _____ = 15,000

5. _____ * 9 = 54

_____ * 90 = 540

_____ * 90 = 5,400

60 * _____ = 540

6 * _____ = 5,400

6 * _____ = 54,000

6. 8 * _____ = 40

8 * _____ = 4,000

80 * _____ = 4,000

_____ * 50 = 400

_____ * 5 = 400

_____ * 500 = 400,000

Practice

7. _____ = 6.3 + 8.7

8. 7.36 + 2.14 = _____

9. _____ = 9.74 − 5.48

10. _____ = 4.6 − 2.8

STUDY LINK 5·3 Estimating Sums

For all problems, write a number model to estimate the sum.

◆ If the estimate is greater than or equal to 1,500, find the exact sum.

◆ If the estimate is less than 1,500, **do not** solve the problem.

1. 867 + 734 = _____

Number model:

_____

2. 374 + 962 + 488 = _____

Number model:

_____

3. 382 + 744 = _____

Number model:

_____

4. 581 + 648 + 366 = _____

Number model:

_____

5. 318 + 295 + 493 = _____

Number model:

_____

6. 845 + 702 = _____

Number model:

_____

7. 694 + 210 + 386 = _____

Number model:

_____

8. 132 + 692 + 803 = _____

Number model:

_____

9. 756 + 381 + 201 = _____

Number model:

_____

10. 575 + 832 = _____

Number model:

_____

Practice

11. 60 * 80 = _____

12. 30 * 70 = _____

13. 50 * 900 = _____

14. 40 * 800 = _____

STUDY LINK 5·4 **Estimating Products**

Estimate whether the answer will be in the tens, hundreds, thousands, or more. Write a number model to show how you estimated. Then circle the box that shows your estimate.

1. A koala sleeps an average of 22 hours each day. About how many hours does a koala sleep in a year?

Number model: _____

10s	100s	1,000s	10,000s	100,000s	1,000,000s

2. A prairie vole (a mouselike rodent) has an average of 9 babies per litter. If it has 17 litters in a season, about how many babies are produced?

Number model: _____

10s	100s	1,000s	10,000s	100,000s	1,000,000s

3. Golfers lose, on average, about 5 golf balls per round of play. About how many golf balls will an average golfer lose playing one round every day for one year?

Number model: _____

10s	100s	1,000s	10,000s	100,000s	1,000,000s

4. In the next hour, the people in France will save 12,000 trees by recycling paper. About how many trees will they save in two days?

Number model: _____

10s	100s	1,000s	10,000s	100,000s	1,000,000s

Try This

5. How many digits can the product of two 2-digit numbers have? Give examples to support your answer.

_____

Practice

6. 60 * 7 = _____   **7.** 4 * 80 = _____   **8.** _____ = 200 * 9

101

STUDY LINK 5·5 Multiplication

Multiply using the partial-product method. Show your work in the grid below.

1. 56 * 7 = _____

2. 8 * 275 = _____

3. _____ = 1,324 * 9

4. Maya goes to school for 7 hours each day. If she does not miss any of the 181 school days, how many hours will Maya spend in school this year?

   a. Estimate whether the answer will be in the tens, hundreds, thousands, or more. Write a number model to show how you estimated. Circle the box that shows your estimate.

   Number model: _____

10s	100s	1,000s	10,000s	100,000s	1,000,000s

   b. Exact answer: _____ hours

5. The average eye blinks once every 5 seconds. Is that more than or less than a hundred thousand times per day? Explain your answer.

_____

_____

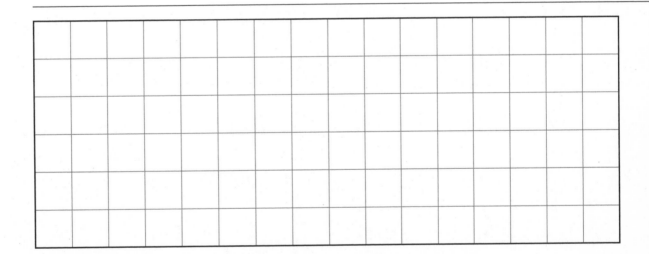

Practice

6. _____ = 495 + 7,389

7. 5,638 + 5,798 = _____

8. 3,007 − 1,749 = _____

9. _____ = 8,561 − 3,872

STUDY LINK 5·6 | More Multiplication

Multiply using the partial-products algorithm. Show your work.

1. 582 * 7 = _____

2. 56 * 30 = _____

3. 42 * 50 = _____

4. _____ = 27 * 18

5. _____ = 46 * 71

6. 340 * 50 = _____

Try This

7. _____ = 241 * 31

8. _____ = 768 * 49

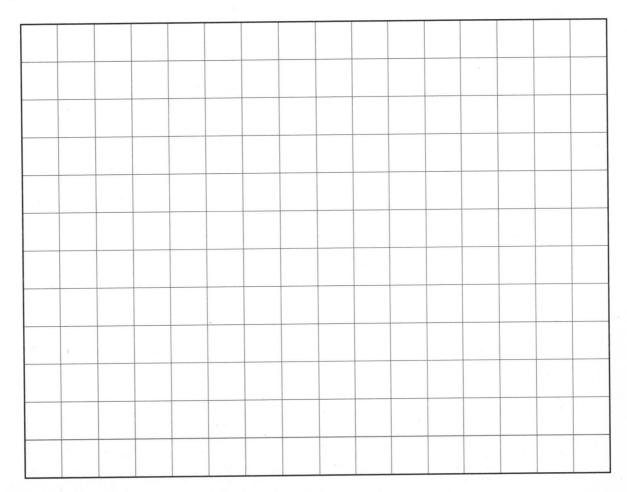

Practice

9. _____ = 283 + 5,439

10. 6,473 + 4,278 = _____

11. 5,583 − 4,667 = _____

12. _____ = 9,141 − 6,372

Lattice Multiplication

Use the lattice method to find the following products.

1. 5 * 46 = _____

2. 8 * 67 = _____

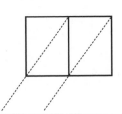

3. 7 * 836 = _____

4. 4 * 329 = _____

5. 25 * 31 = _____

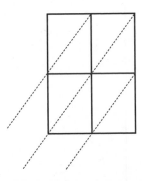

6. 49 * 52 = _____

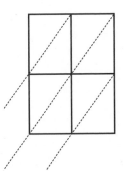

7. Use the lattice method and the partial-products method to find the product.

84 * 78 = _____

Practice

8. _____ = 33.67 + 5.9

9. 68.4 + 5.82 = _____

10. 71.44 − 37.67 = _____

11. _____ = 101.06 − 29.91

STUDY LINK 5·8 Place-Value Puzzle

Use the clues below to fill in the place-value chart.

Billions				Millions				Thousands				Ones		
100B	10B	1B	,	100M	10M	1M	,	100Th	10Th	1Th	,	100	10	1

1. Find $\frac{1}{2}$ of 24. Subtract 4. Write the result in the hundreds place.

2. Find $\frac{1}{2}$ of 30. Divide the result by 3. Write the answer in the ten-thousands place.

3. Find $30 \div 10$. Double the result. Write it in the one-millions place.

4. Divide 12 by 4. Write the answer in the ones place.

5. Find $9 * 8$. Reverse the digits in the result. Divide by 3. Write the answer in the hundred-thousands place.

6. Double 8. Divide the result by 4. Write the answer in the one-thousands place.

7. In the one-billions place, write the even number greater than 0 that has not been used yet.

8. Write the answer to $5 \div 5$ in the hundred-millions place.

9. In the tens place, write the odd number that has not been used yet.

10. Find the sum of all the digits in the chart so far. Divide the result by 5, and write it in the ten-billions place.

11. Write 0 in the empty column whose place value is less than billions.

12. Write the number in words. For example, 17,450,206 could be written as "17 million, 450 thousand, 206."

_____

Practice

13. $74 * 5 =$ _____   14. _____ $= 396 * 8$

15. _____ $= 92 * 18$ 16. $56 * 47 =$ _____

109

STUDY LINK 5·9 Many Names for Powers of 10

Below are different names for powers of 10. Write the names in the appropriate name-collection boxes. Circle the names that do not fit in any of the boxes.

1,000,000	10,000	1,000
100	10	10 [100,000s]
10 [10,000s]	10^6	10 [1,000s]
10^3	10 * 10 * 10 * 10	one thousand
10^5	10 * 10 * 10 * 10 * 10	10 [10s]
10 * 10	ten	10^1
10 [tenths]	10^0	1

1. | **100,000** |
| |
| |
| |

2. | 10^2 |
| |
| |
| |

3. | **1 million** |
| |
| |
| |

4. | **one** |
| |
| |
| |

5. | **10 * 10 * 10** |
| |
| |
| |

6. | 10^4 |
| |
| |
| |

Practice

7. 63 * 7 = _____

8. _____ = 495 * 6

9. _____ = 97 * 53

111

STUDY LINK 5·10 | **Rounding**

1. Round the seating capacities in the table below to the nearest thousand.

Women's National Basketball Association Seating Capacity of Home Courts		
Team	**Seating Capacity**	**Rounded to the Nearest 1,000**
Charlotte Sting	24,042	
Cleveland Rockers	20,562	
Detroit Shock	22,076	
New York Liberty	19,763	
Phoenix Mercury	19,023	
Sacramento Monarchs	17,317	
San Antonio Stars	18,500	
Seattle Storm	17,072	

2. Look at your rounded numbers. Which stadiums have about the same capacity?

_____

3. Round the population figures in the table below to the nearest million.

U.S. Population by Official Census from 1940 to 2000		
Year	**Population**	**Rounded to the Nearest Million**
1940	132,164,569	
1960	179,323,175	
1980	226,542,203	
2000	281,421,906	

Source for both tables: The World Almanac and Book of Facts 2004

Practice

4. _____ = 692 * 6   **5.** _____ = 38 * 21   **6.** 44 * 73 = _____

113

Copyright © Wright Group/McGraw-Hill

STUDY LINK
5·11 Comparing Data

This table shows the number of pounds of fruit produced by the top 10 fruit-producing countries in 2001. Read each of these numbers to a friend or a family member.

Country	Pounds of Fruit
Brazil	77,268,294,000
China	167,046,420,000
France	26,823,740,000
India	118,036,194,000
Iran	28,599,912,000
Italy	44,410,538,000
Mexico	34,549,912,000
Philippines	27,028,556,000
Spain	36,260,392,000
United States	73,148,598,000

1. Which country produced the most fruit?

   _____

2. Which country produced the least fruit?

   _____

3. For each pair, circle the country that produced more fruit.

   a. India   Mexico

   b. United States   Iran

   c. Brazil   Philippines

   d. Spain   Italy

4. Which two countries together produced about as much fruit as India?

   _____

Practice

Estimate the sum. Write a number model.

5. 687 + 935 _____

6. 2,409 + 1,196 + 1,327 _____

7. 11,899 + 35,201 _____

STUDY LINK 5·12
Unit 6: Family Letter

Division; Map Reference Frames; Measures of Angles

The first four lessons and the last lesson of Unit 6 focus on understanding the division operation, developing a method for dividing whole numbers, and solving division number stories.

Though most adults reach for a calculator to do a long-division problem, it is useful to know a paper-and-pencil procedure for computations such as $567 \div 6$ and $15\overline{)235}$. Fortunately, there is a method that is similar to the one most of us learned in school but is much easier to understand and use. This method is called the **partial-quotients method.**

Students have had considerable practice with extended division facts, such as $420 \div 7 = 60$, and questions, such as "About how many 12s are in 150?" Using the partial-quotients method, your child will apply these skills to build partial quotients until the exact quotient and remainder are determined.

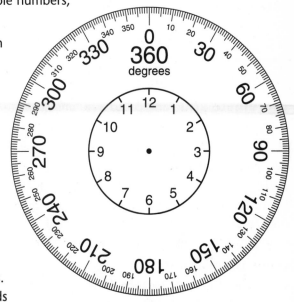

Full-circle (360°) protractor

This unit also focuses on numbers in map coordinate systems. For maps of relatively small areas, rectangular coordinate grids are used. For world maps and the world globe, the system of latitude and longitude is used to locate places.

Because this global system is based on angle measures, the class will practice measuring and drawing angles with full-circle (360°) and half-circle (180°) protractors. If you have a protractor, ask your child to show you how to use this tool.

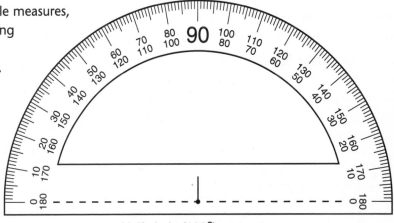

Half-circle (180°) protractor

The class is well into the World Tour. Students have visited Africa and are now traveling in Europe. They are beginning to see how numerical information about a country helps them get a better understanding of the country—its size, climate, location, and population distribution—and how these characteristics affect the way people live. Your child may want to share with you information about some of the countries the class has visited. Encourage your child to take materials about Europe to school, such as magazine articles, travel brochures, and articles in the travel section of your newspaper.

Please keep this Family Letter for reference as your child works through Unit 6.

117

Vocabulary

Important terms in Unit 6:

acute angle An angle with a measure greater than 0° and less than 90°.

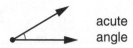

acute angle

coordinate grid (also called a *rectangular coordinate grid*) A reference frame for locating points in a plane using *ordered number pairs,* or *coordinates.*

equal-groups notation A way to denote a number of equal-sized groups. The size of the groups is written inside square brackets and the number of groups is written in front of the brackets. For example, 3 [6s] means 3 groups with 6 in each group.

index of locations A list of places together with a reference frame for locating them on a map. For example, "Billings D3," indicates that Billings can be found within the rectangle where column 3 and row D of a grid meet on the map.

meridian bar A device on a globe that shows degrees of latitude north and south of the equator.

multiplication/division diagram A diagram used for problems in which a total is made up of several equal groups. The diagram has three parts: a number of groups, a number in each group, and a total number.

rows	chairs per row	chairs in all
6	4	24

obtuse angle An angle with a measure greater than 90° and less than 180°.

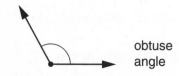

obtuse angle

ordered number pair Two numbers that are used to locate a point on a *coordinate grid.* The first number gives the position along the horizontal axis, and the second number gives the position along the vertical axis. The numbers in an ordered pair are called *coordinates.* Ordered pairs are usually written inside parentheses: (2,3).

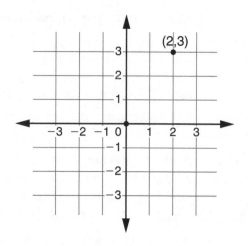

protractor A tool used for measuring or drawing angles. A half-circle protractor can be used to measure and draw angles up to 180°. A full-circle protractor can be used to measure and draw angles up to 360°. One of each type is on the Geometry Template.

quotient The result of dividing one number by another number. For example, in 35 ÷ 5 = 7, the quotient is 7.

reflex angle An angle with a measure greater than 180° and less than 360°.

straight angle An angle with a measure of 180°.

vertex The point at which the rays of an angle, the sides of a polygon, or the edges of a polyhedron meet. Plural is vertexes or vertices.

Do-Anytime Activities

To work with your child on concepts taught in this unit, try these interesting and rewarding activities:

1. Help your child practice division by solving problems for everyday situations.

2. Name places on the world globe and ask your child to give the latitude and longitude for each.

3. Encourage your child to identify and classify acute, right, obtuse, straight, and reflex angles in buildings, bridges, and other structures.

4. Work together with your child to construct a map, coordinate system, and index of locations for your neighborhood.

Building Skills through Games

In Unit 6, your child will practice using division and reference frames and measuring angles by playing the following games. For detailed instructions, see the *Student Reference Book.*

Angle Tangle See *Student Reference Book,* page 230.

This is a game for two players and will require a protractor. The game helps students practice drawing, estimating the measure of, and measuring angles.

Division Dash See *Student Reference Book,* page 241.

This is a game for one or two players. Each player will need a calculator. The game helps students practice division and mental calculation.

Grid Search See *Student Reference Book,* pages 250 and 251.

This is a game for two players, and each player will require *two* playing grids. The game helps students practice using a letter-number coordinate system and developing a search strategy.

Over and Up Squares See *Student Reference Book,* page 257.

This is a game for two players and will require a playing grid. The game helps students practice using ordered pairs of numbers to locate points on a rectangular grid.

As You Help Your Child with Homework

As your child brings assignments home, you may want to go over the instructions together, clarifying them as necessary. The answers listed below will guide you through some of the Study Links in this unit.

Study Link 6·1

1. 8 rows   **2.** 120,000 quills   **3.** 21 boxes

Study Link 6·2

1. 38   **2.** 23   **3.** 47

Study Link 6·3

1. 13 marbles; 5 left   **2.** 72 prizes, 0 left
3. 22 R3   **4.** 53 R3

Study Link 6·4

1. $15\frac{4}{8}$ or $15\frac{1}{2}$; Reported it as a fraction or decimal; Sample answer: You can cut the remaining strawberries into halves to divide them evenly among 8 students.

2. 21; Ignored it; Sample answer: There are not enough remaining pens to form another group of 16.

Study Link 6·5

1–7.

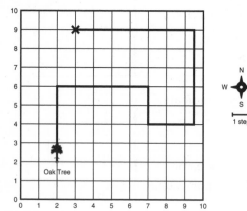

Study Link 6·6

1. >; 101°   **2.** <; 52°
3. >; 144°   **4.** <; 137°
6. 24   **7.** 8 R2   **8.** 157   **9.** 185 R3

Study Link 6·7

1. 60°   **2.** 150°   **3.** 84°   **4.** 105°
5. 32°   **6.** 300°

Study Link 6·8

1.

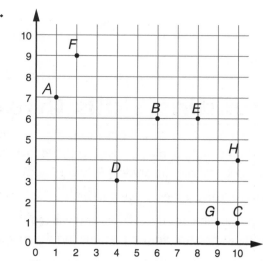

2. K (4,8); L (7,7); M (10,5); N (1,8); O (6,2);
P (8,4); Q (10,2); R (3,10)

Study Link 6·9

1.

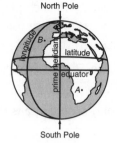

2. Eastern Hemisphere   **3.** water
4. 15 R2   **5.** 14   **6.** 43 R2   **7.** 134

Study Link 6·10

1. 8 pitchers; 0 oranges left

2. 22 bouquets; 8 flowers left

3. 45 R6   **4.** 69   **5.** 180

6. 2,233   **7.** 1,827   **8.** 16,287

Multiplication/Division Number Stories

Fill in each Multiplication/Division Diagram. Then write a number model.
Be sure to include a unit with your answer.

1. Trung wants to rearrange his collection of 72 animals on a shelf in his room.
How many equal rows of 9 animals can he make?

rows	animals per row	animals in all

Number model: _____

Answer: _____

2. An average porcupine has about 30,000 quills. About how many quills
would 4 porcupines have?

porcupines	quills per porcupine	quills in all

Number model: _____

Answer: _____

3. There are 168 calculators for the students at Madison School. A box holds
8 calculators. How many boxes are needed to hold all of the calculators?

boxes	calculators per box	calculators in all

Number model: _____

Answer: _____

Practice

4. _____ $= 6.17 + 8.77$ **5.** _____ $= 12.13 - 4.44$

STUDY LINK 6·2 Equal-Grouping Division Problems

For Problems 1–3, fill in the multiples-of-10 list if it is helpful. If you prefer to solve the division problems in another way, show your work.

1. The community center bought 228 juice boxes for a picnic. How many 6-packs is that?

10 [6s] = _____   Number model: _____

20 [6s] = _____   Answer: _____ 6-packs

30 [6s] = _____

40 [6s] = _____

50 [6s] = _____

2. There are 8 girls on each basketball team. There are 184 girls in the league. How many teams are there?

10 [8s] = _____   Number model: _____

20 [8s] = _____   Answer: _____ teams

30 [8s] = _____

40 [8s] = _____

50 [8s] = _____

3. How many 3s are in 142?

10 [3s] = _____   Number model: _____

20 [3s] = _____   Answer: _____

30 [3s] = _____

40 [3s] = _____

50 [3s] = _____

Practice

4. _____ = 661 * 4   **5.** 13 * 96 = _____   **6.** _____ = 59 * 82

123

STUDY LINK 6·3 Division

1. Bernardo divided a bag of 83 marbles evenly among five friends and himself. How many marbles did each get?

   Number model: _____

   Answer: _____ marbles

   How many marbles are left over?

   _____ marbles

2. The carnival committee has 360 small prizes to share equally with 5 carnival booths. How many prizes will each booth get?

   Number model: _____

   Answer: _____ prizes

   How many prizes are left over?

   _____ prizes

3. 4)‾91   Answer: _____

4. 427 / 8   Answer: _____

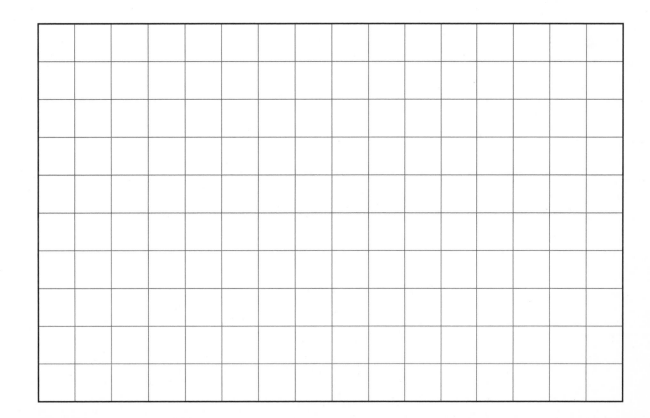

Practice

5. _____ = 34.96 + 1.58

6. _____ = 300.2 + 2.378

7. 43.27 − 12.67 = _____

8. 74.6 − 31.055 = _____

**STUDY LINK
6·4**

Interpreting Remainders

1. Mrs. Patel brought a box of 124 strawberries to the party. She wants to divide the strawberries evenly among 8 people. How many strawberries will each person get?

Picture:

Number model: _____

Answer: _____ strawberries

What did you do about the remainder? Circle the answer.

A. Ignored it

B. Reported it as a fraction or decimal

C. Rounded the answer up

Why? _____

_____

_____

2. Mr. Chew has a box of 348 pens. He asks Maurice to divide the pens into groups of 16. How many groups can Maurice make?

Picture:

Number model: _____

Answer: _____ groups

What did you do about the remainder? Circle the answer.

A. Ignored it

B. Reported it as a fraction or decimal

C. Rounded the answer up

Why? _____

_____

_____

Practice

3. $68 \div 7 =$ _____

4. _____ $= 74 \div 4$

5. $\dfrac{468}{9} =$ _____

6. $3\overline{)95} =$ _____

STUDY LINK 6·5 | Treasure Hunt

Marge and her friends are playing Treasure Hunt. Help them find the treasure.
Follow the directions. Draw the path from the oak tree to the treasure. Mark the
spot where the treasure is buried.

1. Start at the dot under the oak tree; face north. Walk 4 steps.

2. Make a quarter turn, clockwise. Walk 5 steps.

3. Face south. Walk 2 steps.

4. Face east. Walk $2\frac{1}{2}$ steps.

5. Make a $\frac{3}{4}$ turn, clockwise. Walk 5 steps.

6. Make a $\frac{3}{4}$ turn, clockwise. Walk $6\frac{1}{2}$ steps.

7. Make an X to mark the spot where you end.

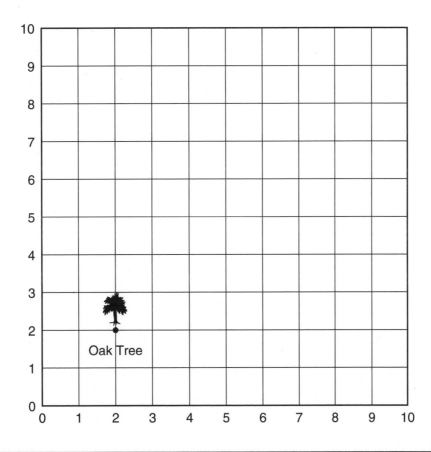

Practice

8. 88 ÷ 3 = _____

9. _____ = 71 ÷ 6

10. _____ = 603 / 7

11. 934 / 5 = _____

129

STUDY LINK 6·6 Measuring Angles

First estimate and then use your full-circle protractor to measure each angle.

1. This angle is _____ (>, <) 90°.

∠G: _____°

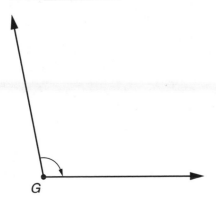

2. This angle is _____ (>, <) 90°.

∠H: _____°

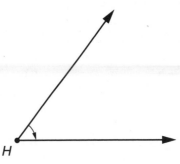

3. This angle is _____ (>, <) 90°.

∠I: _____°

4. This angle is _____ (>, <) 90°.

∠J: _____°

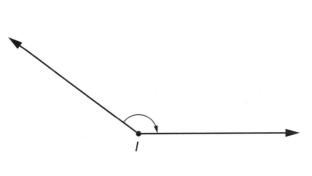

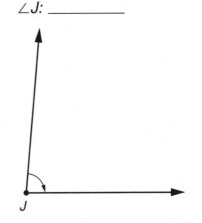

Try This

5. On the back of this page, draw and label angles with the following degree measures:

∠ABC 78°   ∠DEF 145°   ∠GHI 213°   ∠JKL 331°

Practice

6. _____ = 96 ÷ 4

7. 66 ÷ 8 = _____

8. _____ = 314 ÷ 2

9. 928 ÷ 5 = _____

STUDY LINK
6·7

Measuring Angles with a Protractor

First estimate whether the angles measure more or less than 90°. Then use a half-circle protractor to measure them.

1. ∠A: _____

2. ∠B: _____

3. ∠C: _____

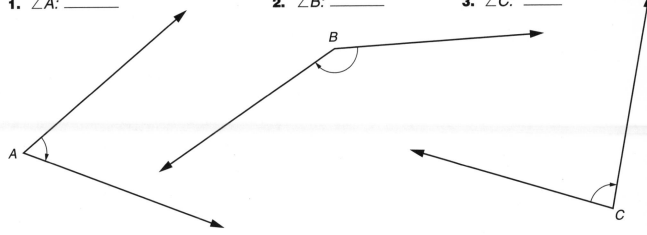

Try This

4. ∠QRS: _____

5. ∠NOP: _____

6. ∠KLM: _____

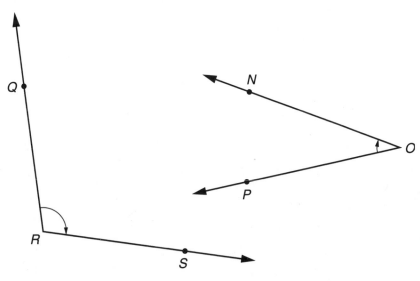

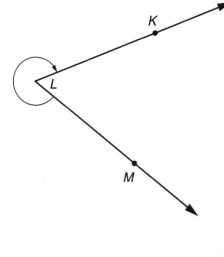

Practice

7. 93 * 6 = _____

8. _____ = 547 * 7

9. _____ = 48 * 39

10. 51 * 64 = _____

133

STUDY LINK 6·8   Coordinate Grids

1. Plot and label each point on the coordinate grid.

A   (1,7)

B   (6,6)

C   (10,1)

D   (4,3)

E   (8,6)

F   (2,9)

G   (9,1)

H   (10,4)

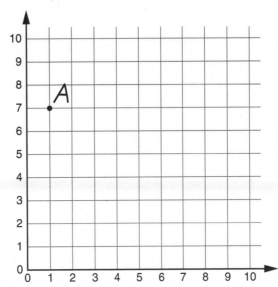

2. Write the ordered number pair for each point plotted on the coordinate grid.

I   (__5__, __3__)

J   (__7__, __2__)

K   (_____, _____)

L   (_____, _____)

M   (_____, _____)

N   (_____, _____)

O   (_____, _____)

P   (_____, _____)

Q   (_____, _____)

R   (_____, _____)

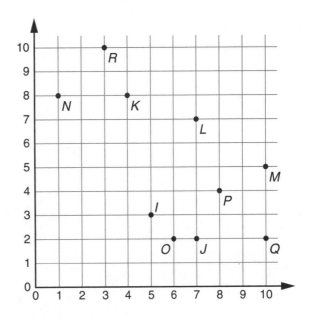

Practice

3. $28 * 7 =$ _____

4. $304 * 5 =$ _____

5. _____ $= 52 * 89$

6. _____ $= 43 * 36$

STUDY LINK 6·9 Latitude and Longitude

Use your *Student Reference Book* to help you complete this Study Link.
Read the examples and study the figures on pages 272 and 273.

SRB 272 273

1. Do the following on the picture of the world globe.

  a. Label the North and South Poles.

  b. Draw and label the equator.

  c. Label the prime meridian.

  d. Draw and label a line of latitude that is north of the equator.

  e. Draw and label a line of longitude that is west of the prime meridian.

  f. Mark a point that is in the Southern Hemisphere and also in the Eastern Hemisphere. Label the point *A*.

  g. Mark a point that is in the Northern Hemisphere and also in the Western Hemisphere. Label the point *B*.

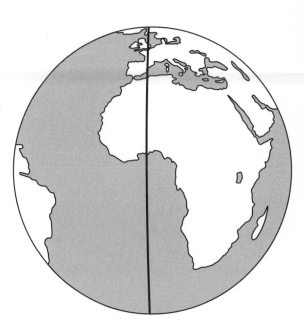

2. The entire continent of Africa is shown in the figure above. Is Africa mostly in the Western Hemisphere or in the Eastern Hemisphere?

_____

3. Do the equator and prime meridian meet over water or over land? _____

Practice

4. _____ = 47 / 3

5. 7)98 _____

6. 217 ÷ 5 = _____

7. _____ = 804 / 6

STUDY LINK 6·10 **Division**

1. It takes 14 oranges to make a small pitcher of juice. Annette has 112 oranges. How many pitchers of juice can she make?

   Number model: _____

   Answer: _____ pitchers of juice

   How many oranges are left over? _____ oranges

2. Each bouquet needs 17 flowers. The florist has 382 flowers in his store. How many bouquets can the florist make?

   Number model: _____

   Answer: _____ bouquets

   How many flowers are left over? _____ flowers

3. 726 ÷ 16 = _____

4. 4)276 _____

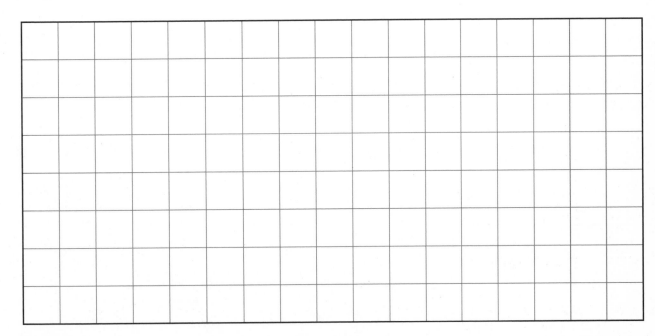

Practice

5. 45 * 4 = _____

6. _____ = 319 * 7

7. _____ = 29 * 63

8. 89 * 183 = _____

139

STUDY LINK 6·11

Unit 7: Family Letter

Fractions and Their Uses; Chance and Probability

One of the most important ideas in mathematics is the concept that a number can be named in many different ways. For example, a store might advertise an item at $\frac{1}{2}$ off its original price or at a 50% discount—both mean the same thing. Much of the mathematics your child will learn involves finding equivalent names for numbers.

A few weeks ago, the class studied decimals as a way of naming numbers between whole numbers. Fractions serve the same purpose. After reviewing the meaning and uses of fractions, students will explore equivalent fractions—fractions that have the same value, such as $\frac{1}{2}, \frac{2}{4}, \frac{3}{6}$, and so on. As in past work with fractions, students will handle concrete objects and look at pictures, because they first need to "see" fractions in order to understand what fractions mean.

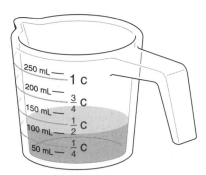

A measuring cup showing fractional increments

Fractions are also used to express the chance that an event will occur. For example, if we flip a coin, we say that it will land heads-up about $\frac{1}{2}$ of the time. The branch of mathematics that deals with chance events is called **probability.** Your child will begin to study probability by performing simple experiments.

Please keep this Family Letter for reference as your child works through Unit 7.

Vocabulary

Important terms in Unit 7:

denominator The number below the line in a fraction. In a fraction where the whole is divided into equal parts, the denominator represents the number of equal parts into which the whole (or ONE or unit whole) is divided. In the fraction $\frac{a}{b}$, b is the denominator.

equal chance outcomes or **equally likely outcomes** If each of the possible outcomes for a chance experiment or situation has the same chance of occurring, the outcomes are said to have an equal chance or to be equally likely. For example, there is an equal chance of getting heads or tails when flipping a coin, so heads and tails are equally likely outcomes.

equivalent fractions Fractions with different denominators that name the same number. For example, $\frac{1}{2}$ and $\frac{4}{8}$ are equivalent fractions.

fair (coin, die, or spinner) A device that is free from bias. Each side of a fair die or coin will come up about equally often. Each section of a fair spinner will come up in proportion to its area.

A die has six faces. If the die is fair, each face has the same chance of coming up.

fair game A game in which every player has the same chance of winning.

mixed number A number that is written using both a whole number and a fraction. For example, $2\frac{1}{4}$ is a mixed number equal to $2 + \frac{1}{4}$.

numerator The number above the line in a fraction. In a fraction where the whole (or ONE or unit whole) is divided into a number of equal parts, the numerator represents the number of equal parts being considered. In the fraction $\frac{a}{b}$, a is the numerator.

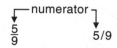

probability A number from 0 through 1 that tells the chance that an event will happen. The closer a probability is to 1, the more likely the event is to happen.

whole (or ONE or unit whole) The entire object, collection of objects, or quantity being considered; the ONE; the unit whole; 100%.

"whole" box In *Everyday Mathematics,* a box in which students write the name of the whole (or ONE or unit whole).

Whole
24 pennies

Do-Anytime Activities

To work with your child on concepts taught in this unit, try these interesting and rewarding activities:

1. Have your child look for everyday uses of fractions in grocery items, clothing sizes, cookbooks, measuring cups and spoons, and statistics in newspapers and on television.

2. Encourage your child to express numbers, quantities, and measures, such as a quarter of an hour, a quart of orange juice, a dozen eggs, and a pint of milk.

3. While grocery shopping, help your child compare prices by looking at shelf labels or calculating unit prices. Help your child make decisions about the "better buy." If a calculator is available, have your child take it to the store.

4. Have your child look for everyday uses of probabilities in games, sports, and weather reports. Ask your child to make a list of events that could never happen, might happen, and are sure to happen.

Building Skills through Games

In this unit, your child will work on his or her understanding of fractions and probability by playing the following games. For detailed instructions, see the *Student Reference Book.*

Chances Are See *Student Reference Book,* pages 236 and 237.
This game is for 2 players and requires one deck of *Chances Are* Event Cards and one deck of *Chances Are* Probability Cards. The game develops skill in using probability terms to describe the likelihood of events.

Fraction Match See *Student Reference Book,* page 243.
This game is for 2 to 4 players and requires one deck of *Fraction Match* cards. The game develops skill in naming equivalent fractions.

Fraction Of See *Student Reference Book,* pages 244 and 245.
This game is for 2 players and requires one deck of *Fraction Of* Fraction Cards and one deck of *Fraction Of* Set Cards. The game develops skill in finding the fraction of a number.

Fraction Top-It See *Student Reference Book,* page 247.
This is a game for 2 to 4 players and requires one set of 32 Fraction Cards. The game develops skill in comparing fractions.

Getting to One See *Student Reference Book,* page 248.
This is a game for 2 players and requires one calculator. The game develops skill in estimation.

Grab Bag See *Student Reference Book,* page 249.
This game is for 2 players or two teams of 2 and requires one deck of *Grab Bag* cards. The game develops skill in calculating the probability of an event.

As You Help Your Child with Homework

As your child brings assignments home, you may want to go over the instructions together, clarifying them as necessary. The answers listed below will guide you through some of the Study Links in this unit.

Study Link 7·2

1. **b.** 4 **c.** 12 **d.** 8   2. 6
3. 12   4. 7   5. 28
6. 10   7. 30   8. 10
9. 12   10. 12   11. $2\frac{1}{2}$
12. 23   13. $19\frac{2}{3}$ 14. 13
15. $41\frac{7}{9}$

Study Link 7·3

1. 50-50 chance   2. very unlikely
4. 5   5. 592   6. 3,948
7. 1,690   8. 16,170

Study Link 7·4

3. 8   4. 0.881   5. 9.845
6. 1.59   7. 0.028

Study Link 7·5

1. Less than $1.00; 0.75 + 0.10 = 0.85
2. $3\frac{3}{4}$ 3. $\frac{1}{6}$ 4. $2\frac{3}{8}$
5. Sample answers:

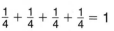
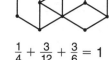

$\frac{1}{4} + \frac{1}{4} + \frac{1}{4} + \frac{1}{4} = 1$ $\frac{1}{4} + \frac{3}{12} + \frac{3}{6} = 1$

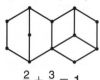

$\frac{2}{4} + \frac{3}{6} = 1$

6. 8   7. 45   8. 49   9. 22

Study Link 7·6

1. C, F, I   2. B, D   3. E, H   4. A, G
5. $\frac{2}{3}$ 7. $\frac{5}{6}$ 9. $\frac{1}{2}$ 10. $\frac{1}{6}$

Study Link 7·7

5. $23\frac{3}{4}$ 6. 19   7. 42

Study Link 7·8

Sample answers for 1–10:

1. $\frac{2}{10}$; $\frac{1}{5}$; $\frac{20}{100}$ 2. $\frac{6}{10}$; $\frac{3}{5}$; $\frac{60}{100}$

3. $\frac{5}{10}$; $\frac{1}{2}$; $\frac{50}{100}$ 4. $\frac{3}{4}$; $\frac{30}{40}$; $\frac{75}{100}$

5. 0.3   6. 0.63   7. 0.7   8. 0.4
9. 0.70; $\frac{70}{100}$ 10. 0.2; $\frac{2}{10}$ 11. 702 12. 3,227
13. 975

Study Link 7·9

1. >   2. <   3. =
4. =   5. <   6. >
7. Answers vary.   8. Answers vary.
9. $\frac{1}{4}$; $\frac{4}{10}$; $\frac{3}{7}$; $\frac{24}{50}$ 10. $\frac{1}{12}$; $\frac{3}{12}$; $\frac{7}{12}$; $\frac{8}{12}$; $\frac{11}{12}$

11. $\frac{1}{50}$; $\frac{1}{20}$; $\frac{1}{5}$; $\frac{1}{3}$; $\frac{1}{2}$ 12. $\frac{4}{100}$; $\frac{4}{12}$; $\frac{4}{8}$; $\frac{4}{5}$; $\frac{4}{4}$

13. 5   14. 100   15. 36

Study Link 7·10

3. 28   4. 27   5. 30   6. 36

Study Link 7·11

3. 29   4. $16\frac{1}{2}$ 5. 105   6. $141\frac{1}{5}$

Study Link 7·12

1. Answers vary.
2. Answers vary.
3. Answers vary.
4. **a.** $\frac{1}{4}$ **b.** $\frac{1}{4}$ **c.** $\frac{1}{2}$
5. Sample answer: I think it will be about the same fraction for 1000 times as it was for 20.
6. 336   7. 7,866   8. 3,870   9. 4,828

STUDY LINK 7·1 Fractions

1. Divide the circle into 6 equal parts.
Color $\frac{5}{6}$ of the circle.

Whole
circle

2. Divide the rectangle into 3 equal parts.
Shade $\frac{2}{3}$ of the rectangle.

Whole
rectangle

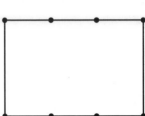

3. Divide each square into fourths.
Color $1\frac{3}{4}$ of the squares.

Whole
square

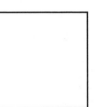

Fill in the missing fractions and mixed numbers on the number lines.

4.

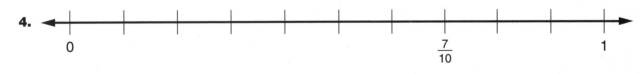

0                       $\frac{7}{10}$ 1

5.

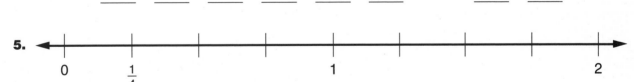

0     $\frac{1}{4}$ 1   2

Practice

6. $854 + 267 =$ _____

7. _____ $= 3{,}398 + 2{,}635$

8. _____ $= 6{,}374 - 755$

9. $5{,}947 - 3{,}972 =$ _____

145

STUDY LINK 7·2 "Fraction-of" Problems

1. Theresa had 24 cookies. She gave $\frac{1}{6}$ to her sister and $\frac{3}{6}$ to her mother.

Whole

   a. Fill in the "whole" box.

   b. How many cookies did she give to her sister? _____ cookies

   c. How many did she give to her mother? _____ cookies

   d. How many did she have left? _____ cookies

Solve.

2. $\frac{1}{3}$ of 18 = _____

3. $\frac{2}{3}$ of 18 = _____

4. $\frac{1}{5}$ of 35 = _____

5. $\frac{4}{5}$ of 35 = _____

6. $\frac{1}{4}$ of 40 = _____

7. $\frac{3}{4}$ of 40 = _____

Try This

8. $\frac{5}{8}$ of 16 = _____

9. $\frac{4}{9}$ of 27 = _____

10. $\frac{3}{5}$ of 20 = _____

11. What is $\frac{1}{4}$ of 10? _____ Explain. _____

_____

_____

Practice

12. 92 ÷ 4 = _____

13. 59 / 3 = _____

14. _____ = 104 / 8

15. $9\overline{)376}$ = _____

Color Tiles

SRB
45 80

There are 5 blue, 2 red, 1 yellow, and 2 green tiles in a bag.

1. Without looking, Maren picks a tile from the bag. Which of these best describes her chances of picking a blue tile?

Ⓐ likely

Ⓑ 50-50 chance

Ⓒ unlikely

Ⓓ very unlikely

2. Which of these best describes her chances of picking a yellow tile?

Ⓐ certain

Ⓑ likely

Ⓒ 50-50 chance

Ⓓ very unlikely

3. Find the probability of each event. Then make up an event and find the probability.

Event	Favorable Outcomes	Possible Outcomes	Probability
Pick a blue tile	5	10	$\frac{5}{10}$
Pick a red tile		10	$\frac{\square}{10}$
Pick a yellow tile		10	$\frac{\square}{10}$
Pick a green tile		10	$\frac{\square}{10}$
Pick a blue, red, or green tile		10	$\frac{\square}{10}$
		10	$\frac{\square}{10}$

4. Suppose you picked a color tile from the bag 10 times. After each pick, you put the tile back in the bag. How many times would you expect to pick a blue tile? _____ times

Try the experiment. Compare your prediction with the actual results.

Practice

5. $74 * 8 =$ _____

6. _____ $= 4 * 987$

7. _____ $= 65 * 26$

8. $35 * 462 =$ _____

STUDY LINK 7·4 Dimensioning Squares

Dividing Squares

Use a straightedge and the dots below to help you divide each of the squares into equal parts.

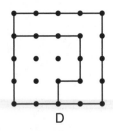

Example: Squares A, B, C, and D are each divided in half in a different way.

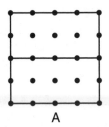 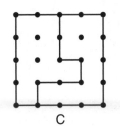 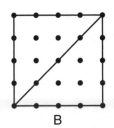

A   B   C   D

1. Square E is divided into fourths. Divide squares F, G, and H into fourths, each in a different way.

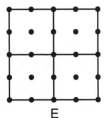 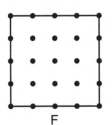 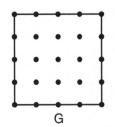 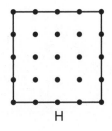

E   F   G   H

2. Square I is divided into eighths. Divide squares J, K, and L into eighths, each in a different way.

 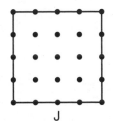 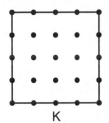 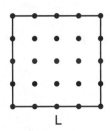

I   J   K   L

3. Rosa has 15 quarters and 10 nickels. She buys juice from a store for herself and her friends. The juice costs 35 cents per can. She gives the cashier $\frac{2}{3}$ of the quarters and $\frac{3}{5}$ of the nickels. The cashier does not give her any change.

How many cans of juice did she buy? _____ cans

Show your work on the back of this paper.

Practice

4. $0.636 + 0.245 =$ _____

5. _____ $= 9.085 + 0.76$

6. _____ $= 1.73 - 0.14$

7. $0.325 - 0.297 =$ _____

151

STUDY LINK 7·5 Fractions

SRB 55 57

1. Jake has $\frac{3}{4}$ of a dollar. Maxwell has $\frac{1}{10}$ of a dollar.
   Do they have more or less than $1.00 in all? _____

   Number model: _____

2. Jillian draws a line segment $2\frac{1}{4}$ inches long. Then she makes the
   line segment $1\frac{1}{2}$ inches longer. How long is the line segment now? _____ inches

   $2\frac{1}{4}$ in. $1\frac{1}{2}$ in.

3. A pizza was cut into 6 slices. Benjamin ate
   $\frac{1}{3}$ of the pizza and Dana ate $\frac{1}{2}$. What fraction
   of the pizza was left? _____

4. Rafael drew a line segment
   $2\frac{7}{8}$ inches long. Then he erased
   $\frac{1}{2}$ inch. How long is the line
   segment now? _____ inches

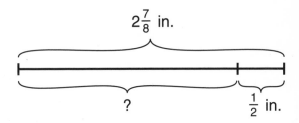

   $2\frac{7}{8}$ in.

   ? $\frac{1}{2}$ in.

5. Two hexagons together are one whole. Draw line segments to divide each
   whole into trapezoids, rhombuses, and triangles. Write a number model
   to show how the parts add up to the whole.

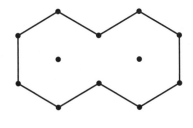 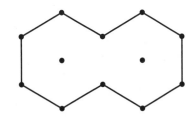 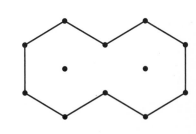

   _____   _____   _____

Practice

6. $\frac{1}{4}$ of 32 = _____   7. _____ = $\frac{9}{10}$ of 50   8. $\frac{7}{8}$ of 56 = _____   9. _____ = $\frac{11}{12}$ of 24

STUDY LINK 7·6   **Many Names for Fractions**

Write the letters of the pictures that represent each fraction.

1. $\frac{1}{2}$ $C,$ _____   **2.** $\frac{3}{4}$ _____

3. $\frac{4}{5}$ _____   **4.** $\frac{2}{3}$ _____

A	**B**	**C**

   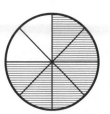   

D	**E**	**F**

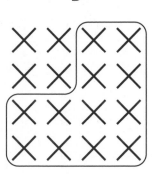   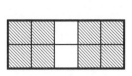   

G	**H**	**I**

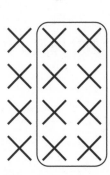      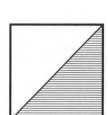

Practice

5. _____ $= \frac{1}{6} + \frac{3}{6}$ **6.** $\frac{2}{4} + \frac{1}{4} =$ _____   **7.** $\frac{1}{2} + \frac{2}{6} =$ _____

8. $\frac{5}{6} - \frac{2}{6} =$ _____   **9.** $\frac{3}{4} - \frac{1}{4} =$ _____   **10.** $\frac{1}{3} - \frac{1}{6} =$ _____

STUDY LINK
7·7

Fraction Name-Collection Boxes

In each name-collection box:

Write the missing number in each fraction so that the fraction belongs in the box. Write one more fraction that can go in the box.

1.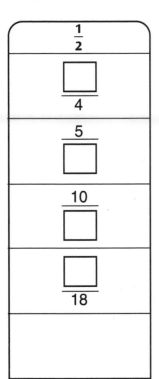

2.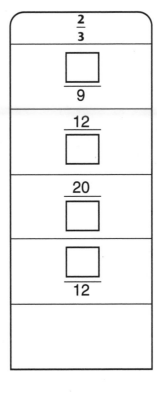

3.

$\frac{1}{4}$
$\frac{\square}{12}$
$\frac{5}{\square}$
$\frac{10}{\square}$
$\frac{\square}{100}$

4. Make up your own name-collection box problems like the ones above. Ask a friend to solve your problems. Check your friend's work.

a.

b.

Practice

5. _____ = 95 / 4   6. 57 ÷ 3 = _____   7. _____ = 882 / 21

157

STUDY LINK 7·8 Fractions and Decimals

Write 3 equivalent fractions for each decimal.

Example:

| 0.8 | $\frac{8}{10}$ | $\frac{4}{5}$ | $\frac{80}{100}$ |

1. 0.20 _____ _____ _____

2. 0.6 _____ _____ _____

3. 0.50 _____ _____ _____

4. 0.75 _____ _____ _____

Write an equivalent decimal for each fraction.

5. $\frac{3}{10}$ _____   **6.** $\frac{63}{100}$ _____   **7.** $\frac{7}{10}$ _____   **8.** $\frac{2}{5}$ _____

9. Shade more than $\frac{53}{100}$ of the square and less than $\frac{8}{10}$ of the square. Write the value of the shaded part as a decimal and a fraction.

Decimal: _____

Fraction: _____

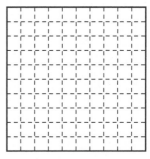

10. Shade more than $\frac{11}{100}$ of the square and less than $\frac{1}{4}$ of the square. Write the value of the shaded part as a decimal and a fraction.

Decimal: _____

Fraction: _____

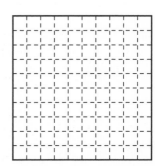

Practice

11. _____ = 78 * 9   **12.** 461 * 7 = _____   **13.** _____ = 39 * 25

159

STUDY LINK
7·9

Compare and Order Fractions

Write <, >, or = to make each number sentence true.

1. $\frac{5}{6}$ _____ $\frac{1}{6}$

2. $\frac{3}{10}$ _____ $\frac{3}{4}$

3. $\frac{2}{3}$ _____ $\frac{10}{15}$

4. $\frac{10}{40}$ _____ $\frac{4}{16}$

5. $\frac{4}{9}$ _____ $\frac{7}{9}$

6. $\frac{5}{6}$ _____ $\frac{5}{8}$

7. Explain how you solved Problem 1. _____

_____

8. Explain how you solved Problem 2. _____

_____

9. Circle each fraction that is less than $\frac{1}{2}$.

$\frac{7}{8}$ $\frac{1}{4}$ $\frac{4}{10}$ $\frac{7}{12}$ $\frac{5}{9}$ $\frac{3}{7}$ $\frac{24}{50}$ $\frac{67}{100}$

Write the fractions in order from smallest to largest.

10. $\frac{3}{12},$ $\frac{7}{12},$ $\frac{1}{12},$ $\frac{11}{12},$ $\frac{8}{12}$

_____ _____ _____ _____ _____
smallest   **largest**

11. $\frac{1}{5},$ $\frac{1}{3},$ $\frac{1}{20},$ $\frac{1}{2},$ $\frac{1}{50}$

_____ _____ _____ _____ _____
smallest   **largest**

12. $\frac{4}{5},$ $\frac{4}{100},$ $\frac{4}{4},$ $\frac{4}{8},$ $\frac{4}{12}$

_____ _____ _____ _____ _____
smallest   **largest**

Practice

13. $\frac{1}{6}$ of 30 = _____

14. $\frac{3}{4}$ of _____ = 75

15. $\frac{4}{5}$ of 45 = _____

STUDY LINK 7·10 — What Is the ONE?

SRB
44

For Problems 1 and 2, use your Geometry Template or sketch the shapes.

1. Suppose ▢ is $\frac{1}{4}$. Draw each of the following:

Example: $\frac{3}{4}$　　**a.** 1　　**b.** $1\frac{1}{2}$　　**c.** 2

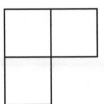

2. Suppose ◇ is $\frac{2}{3}$. Draw each of the following:

a. $\frac{1}{3}$　　**b.** 1　　**c.** $\frac{4}{3}$　　**d.** 2

Use counters to solve the following problems.

3. If 14 counters are $\frac{1}{2}$, then what is the ONE?

_____ counters

4. If 9 counters are $\frac{1}{3}$, then what is the ONE?

_____ counters

5. If 12 counters are $\frac{2}{5}$, then what is the ONE? _____ counters

6. If 16 counters are $\frac{4}{9}$, then what is the ONE? _____ counters

---

Practice

7. _____ $= \frac{1}{4} + \frac{1}{2}$　　**8.** $\frac{1}{3} + \frac{1}{6} =$ _____

9. $\frac{3}{4} - \frac{1}{4} =$ _____　　**10.** _____ $= \frac{5}{6} - \frac{1}{3}$

163

Spinners and Fractions

1. Design your own spinner with as many colors as you wish. Use a pencil until you are satisfied with your work, then color your spinner.

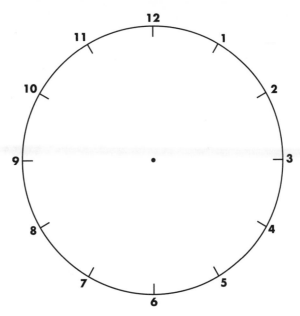

2. Describe your spinner.

    a. The chances of the paper clip landing on _____ are _____ out of _____.
    (color)

    b. The paper clip has a _____ chance of landing on _____.
    (color)

    c. It is unlikely that the paper clip will land on _____.
    (color)

    d. It is _____ times as likely to land on _____ as on _____.
    (color)   (color)

    e. It is more likely to land on _____ than _____.
    (color)   (color)

Practice

3. _____ = 87 ÷ 3

4. 6)̄99 = _____

5. 945 / 9 = _____

6. 706 ÷ 5 = _____

STUDY LINK
7·11

Layout of a Kitchen

**Pages 235 and 236 will be needed to do Lesson 8-1 in the next unit.
Please complete the pages and return them to class.**

Every kitchen needs a stove, a sink, and a refrigerator. Notice how the stove,
sink, and refrigerator are arranged in the kitchen below. The triangle shows
the work path in the kitchen. Walking from the stove to the sink and to the
refrigerator forms an invisible "triangle" on the floor.

Front View of Kitchen

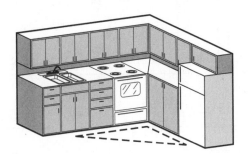

Bird's-Eye View of Kitchen
(looking down at appliances
and countertops)

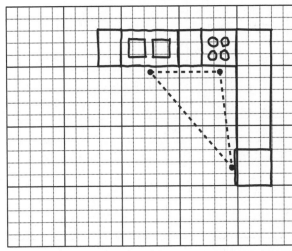

The side of a grid square represents 1 foot.

1. Put one coin or other marker on the floor in front of your sink, one in front
   of your stove, and one in front of your refrigerator.

2. Measure the distance between each pair of markers. Use feet and inches,
   and record your measurements below.

   Distance between

   a. stove and refrigerator   About _____ feet _____ inches

   b. refrigerator and sink   About _____ feet _____ inches

   c. sink and stove   About _____ feet _____ inches

167

STUDY LINK 7·11

Layout of a Kitchen *continued*

3. On the grid below, make a sketch that shows how the stove, sink, and refrigerator are arranged in your kitchen.

Your sketch should show a bird's-eye view of these 3 appliances (including all countertops).

If your oven is separate from your stove, sketch the stove top only.

Use the following symbols in your sketch:

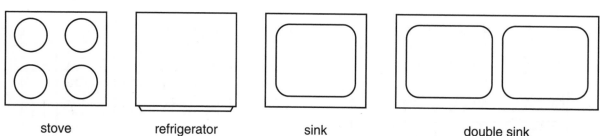

| stove | refrigerator | sink | double sink |

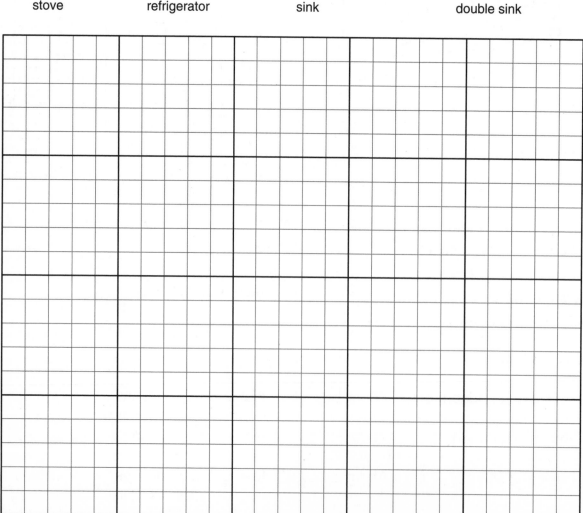

STUDY LINK
7·12

What Are the Chances?

SRB
81

1. You are going to toss 2 pennies 20 times. How many times do you expect the 2 pennies will come up as

   a. 2 heads? _____ times   **b.** 2 tails? _____ times

   c. 1 head and 1 tail? _____ times

2. Now toss 2 pennies together 20 times. Record the results in the table.

A Penny Toss	
Results	**Number of Times**
2 heads	
2 tails	
1 head and 1 tail	

3. What fraction of the tosses came up as

   a. 2 heads? _____   **b.** 2 tails? _____   **c.** 1 head and 1 tail? _____

4. Suppose you were to flip the coins 1,000 times. What fraction do you expect would come up as

   a. 2 heads? _____   **b.** 2 tails? _____

   c. 1 head and 1 tail? _____

5. Explain how you got your answers for Problem 4.

   _____

   _____

   _____

Practice

6. 7 * 48 = _____   7. 874 * 9 = _____

8. _____ = 45 * 86   9. _____ = 34 * 142

169

STUDY LINK 7·13 | **Unit 8: Family Letter**

Perimeter and Area

In previous grades, your child studied the *perimeter* (distance around) and the *area* (amount of surface) of various geometric figures. This next unit will extend your child's understanding of geometry by developing and applying formulas for the areas of figures such as rectangles, parallelograms, and triangles.

Area of a Rectangle

Area = base * height (or length * width)

$A = b * h$ (or $l * w$)

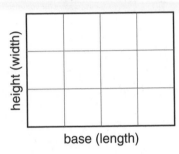

base (length)

Area of a Parallelogram

Area = base * height

$A = b * h$

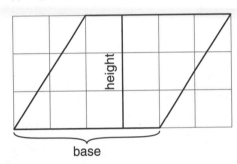

base

Area of a Triangle

Area = $\frac{1}{2}$ of (base * height)

$A = \frac{1}{2} * b * h$

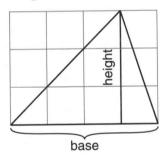

base

Students will learn how to make scale drawings and apply their knowledge of perimeter, area, and scale drawing by analyzing the arrangement of the appliances in their kitchens and the furniture in their bedrooms.

Students will also calculate the area of the skin that covers their entire body. A rule of thumb is that the area of a person's skin is about 100 times the area of one side of that person's hand. Ask your child to show you how to calculate the area of your own skin.

The World Tour will continue. Students will examine how geographical areas are measured and the difficulties in making accurate measurements. They will compare areas for South American countries by using division to calculate the ratio of areas.

Please keep this Family Letter for reference as your child works through Unit 8.

Vocabulary

Important terms in Unit 8:

area The amount of surface inside a closed 2-dimensional (flat) boundary. Area is measured in *square units,* such as square inches or square centimeters.

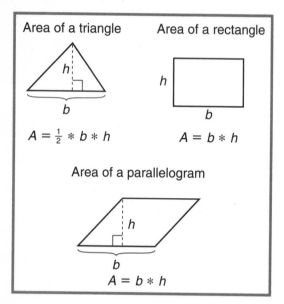

Area of a triangle

$$A = \frac{1}{2} * b * h$$

Area of a rectangle

$$A = b * h$$

Area of a parallelogram

$$A = b * h$$

formula A general rule for finding the value of something. A formula is often written using letter *variables,* which stand for the quantities involved.

length The distance between two points on a 1-dimensional figure. Length is measured in units such as inches, meters, and miles.

perimeter The distance around a 2-dimensional shape along the boundary of the shape. The perimeter of a circle is called its circumference. The perimeter of a polygon is the sum of the lengths of its sides.

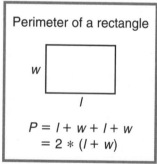

Perimeter of a rectangle

$$P = l + w + l + w$$
$$= 2 * (l + w)$$

perpendicular Crossing or meeting at right angles. Lines, rays, line segments, and planes that cross or meet at right angles are perpendicular. The symbol ⊥ means "is perpendicular to," as in "line *CD* ⊥ line *AB*." The symbol ⊥ indicates a right angle.

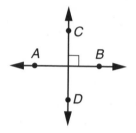

Perpendicular lines

scale The ratio of the distance on a map, globe, drawing, or model to an actual distance.

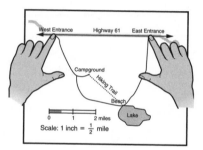

scale drawing A drawing of an object or a region in which all parts are drawn to the same scale as the object. Architects and builders often use scale drawings.

square unit A unit used to measure area. For example, a square that measures one inch on each side has an area of one square inch.

variable A letter or other symbol that represents a number. A variable can represent one specific number, or it can stand for many different numbers.

width The length of one side of a rectangle or rectangular object, typically the shorter side.

Do-Anytime Activities

To work with your child on concepts taught in this unit, try these interesting and engaging activities:

1. Have your child pretend that he or she is a carpenter whose job is to redesign a room—for example, a bedroom, the kitchen, or the living room. Have him or her make a rough estimate of the area of the room. Then help your child check the estimate by finding the actual area using a tape measure or, if possible, blueprints.

2. Have your child pretend that he or she is an architect. Give him or her some dimensions and space requirements to work with. Then have your child design a "dream house," "dream bedroom," or sports stadium, and make a scale drawing for that design.

3. Work with your child to make a scale drawing of your neighborhood. Or have your child make a scale drawing of the floor plan of your house or apartment.

4. Have your child compare the areas of continents, countries, states, or major cities.

Building Skills through Games

In this unit, your child will calculate perimeter and area, compare fractions, identify equivalent fractions, find fractions of collections, and calculate expected probabilities by playing the following games. For detailed instructions, see the *Student Reference Book.*

Fraction Match See *Student Reference Book,* page 243.
This is a game for 2 to 4 players and requires a deck of *Fraction Match* Cards. The game provides practice recognizing equivalent fractions.

Fraction Of See *Student Reference Book,* pages 244 and 245.
This is a game for 2 players and requires 1 deck of *Fraction Of* Fraction Cards, 1 deck of *Fraction Of* Set Cards, and 1 *Fraction Of* Gameboard and Record Sheet. The game provides practice finding fractions of collections.

Fraction Top-It See *Student Reference Book,* page 247.
This is a game for 2 to 4 players and requires a set of Fraction Cards 1 and 2. The game provides practice comparing fractions.

Grab Bag See *Student Reference Book,* page 249.
This is a game for 2 players or two teams of 2 players. It requires 1 deck of *Grab Bag* Cards, 2 *Grab Bag* Record Sheets, and 3 six-sided dice. The game provides practice with variable substitution and calculating probabilities of events.

Rugs and Fences See *Student Reference Book,* pages 260 and 261.
This is a game for 2 players and requires a *Rugs and Fences* Polygon Deck and an Area and Perimeter Deck. The game provides practice finding and comparing the area and perimeter of polygons.

As You Help Your Child with Homework

As your child brings assignments home, you may want to go over the instructions together, clarifying them as necessary. The answers listed below will guide you through some of the Study Links in this unit.

Study Link 8·1

1. 17 feet   2. 54 inches

3.   Sample answer:

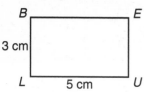

4.   Sample answer:

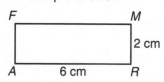

5.

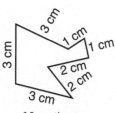

15 centimeters

6.

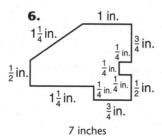

7 inches

Study Link 8·2

1. **a.** 52 miles   **b.** 117 miles

   c. $32\frac{1}{2}$ miles   **d.** $175\frac{1}{2}$ miles

3.

Rectangle	Height in Drawing	Actual Height
A	$\frac{1}{2}$ in.	12 ft
B	$1\frac{1}{4}$ in.	30 ft
C	2 in.	48 ft
D	$1\frac{3}{4}$ in.	42 ft
E	1 in.	24 ft

Study Link 8·3

1. 24 square centimeters

2. 24 square centimeters

2., continued Sample answer:

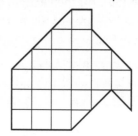

3. 2,072   4. 11,740   5. 3,593   6. 2,848

Study Link 8·4

1. 87,500; 35 grid squares

2. 17,500; 7 grid squares

3. 88.71   4. 58.08   5. 386.174   6. 18.098

Study Link 8·5

1. 48 square feet   2. 21 square inches

3. 864 square centimeters

4. 300 square meters

5. 9 inches   6. 10 centimeters

7. 9, 15, 18, 21   8. 28, 35, 49, 56

9. 36, 54, 60, 66   10. 24, 48, 72, 84

Study Link 8·6

1. $9 * 4 = 36$ 2. $3 * 8 = 24$

3. $4 * 6 = 24$ 4. $65 * 72 = 4,680$

5. 13 inches   6. 85 meters

Study Link 8·7

1. $\frac{1}{2} * (8 * 4) = 16$ 2. $\frac{1}{2} * (12 * 5) = 30$

3. $\frac{1}{2} * (10 * 2) = 10$

4. $\frac{1}{2} * (34 * 75) = 1,275$

5. 3 inches   6. 6 meters

7. 27, 36, 54, 72   8. 8, 24, 40, 48

174

STUDY LINK 8·1 **Perimeter**

1. Perimeter = _____ feet

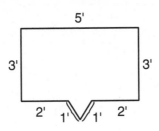

2. Perimeter = _____ inches

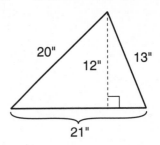

3. Draw a rectangle *BLUE* whose perimeter is 16 centimeters. Label the length of the sides.

4. Draw a different rectangle *FARM* whose perimeter is also 16 centimeters. Label the length of its sides.

5. Measure the sides of the figure to the nearest centimeter. Label the length of its sides. Find its perimeter.

Perimeter = _____ centimeters

6. Measure the sides of the figure to the nearest $\frac{1}{4}$ inch. Label the length of its sides. Find its perimeter.

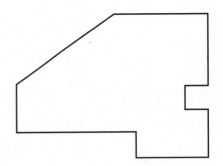

Perimeter = _____ inches

Practice

7. $\frac{1}{4}$ of 24 = _____

8. _____ = $\frac{2}{3}$ of 24

9. _____ = $\frac{5}{8}$ of 40

175

STUDY LINK 8·2 | Scale

SRB
145

1. If 1 inch on a map represents 13 miles, then

a. 4 inches represent _____ miles.

b. 9 inches represent _____ miles.

c. $2\frac{1}{2}$ inches represent _____ miles.

d. $13\frac{1}{2}$ inches represent _____ miles.

2. The scale for a drawing is 1 centimeter : 5 meters. Make a scale drawing of a rectangle that measures 20 meters by 15 meters.

Try This

3. Scale: $\frac{1}{4}$ inch represents 6 feet. Measure the height of each rectangle to the nearest $\frac{1}{4}$ inch. Complete the table.

A

B

C

D

E

Rectangle	Height in Drawing	Actual Height
A		
B		
C		
D		
E		

STUDY LINK 8·3 | Exploring Area

1. Rectangle A at the right is drawn on a 1-centimeter grid. Find its area.

Area = _____ cm²

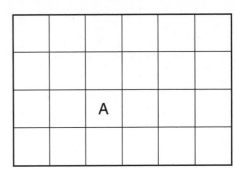

2. Rectangle B has the same area as Rectangle A. Cut out Rectangle B. Then cut it into 5 pieces any way you want.

Rearrange the pieces into a new shape that is not a rectangle. Then tape the pieces together in the space below. What is the area of the new shape?

Area of new shape = _____ cm²

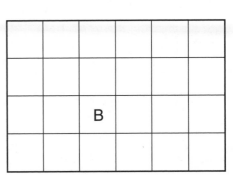

Practice

3. 1,778 + 294 = _____

4. _____ = 6,096 + 5,644

5. 4,007 − 414 = _____

6. _____ = 8,030 − 5,182

STUDY LINK 8·4 | Areas of Irregular Figures

1. Below is a map of São Paulo State in Brazil. Each grid square represents 2,500 square miles. Estimate the area of São Paulo State.

I counted about _____ grid squares.

The area is about _____ square miles.

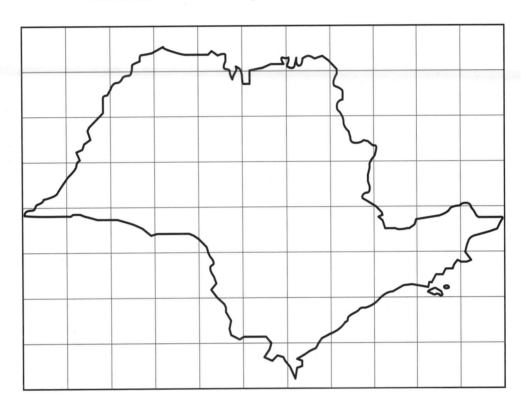

2. To the right is a map of Rio de Janeiro State in Brazil. Each grid square represents 2,500 square miles. Estimate the area of Rio de Janeiro State.

I counted about _____ grid squares.

The area is about _____ square miles.

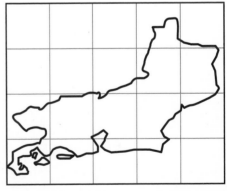

Practice

3. _____ = 73.04 + 15.67

4. 86.05 − 27.97 = _____

5. _____ = 312.11 + 74.064

6. 57.1 − 39.002 = _____

181

STUDY LINK
8·5   **Areas of Rectangles**

Find the area of each rectangle.

1.

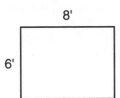
8'

6'

Number model: _____

Area = _____ square feet

2.

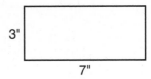
3"

7"

Number model: _____

Area = _____ square inches

3.

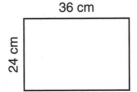
36 cm

24 cm

Number model: _____

Area = _____ square centimeters

4.

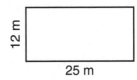
12 m

25 m

Number model: _____

Area = _____ square meters

Try This

The area of each rectangle is given. Find the missing length.

5.

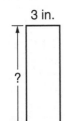
3 in.

?

Area = 27 in^2

height = _____ in.

6.

?

12 cm

Area = 120 cm^2

base = _____ cm

Practice

7. 3, 6, _____, 12, _____, _____, _____

8. 14, 21, _____, _____, 42, _____, _____

9. 30, _____, 42, 48, _____, _____, _____

10. 12, _____, 36, _____, 60, _____, _____

STUDY LINK 8·6 Areas of Parallelograms

Find the area of each parallelogram.

1.

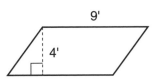

Number model: _____

Area = _____ square feet

2.

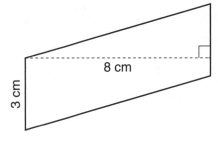

Number model: _____

Area = _____ square centimeters

3.

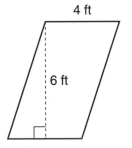

Number model: _____

Area = _____ square feet

4.

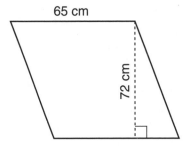

Number model: _____

Area = _____ square centimeters

Try This

The area of each parallelogram is given. Find the length of the base.

5.

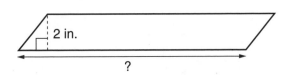

Area = 26 square inches

base = _____ inches

6.

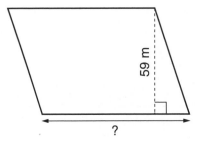

Area = 5,015 square meters

base = _____ meters

STUDY LINK 8·6 — Percents in My World

Percent means "per hundred" or "out of a hundred." *1 percent* means $\frac{1}{100}$ or 0.01.

"48 percent of the students in our school are boys" means that out of every 100 students in the school, 48 are boys.

Percents are written in two ways: with the word *percent,* as in the sentence above, and with the symbol %.

Collect examples of percents. Look in newspapers, magazines, books, almanacs, and encyclopedias. Ask people at home to help. Write the examples below. Also tell where you found them. If an adult says you may, cut out examples and bring them to school.

Encyclopedia: 91% of the area of New Jersey is land, and 9% is covered by water.

Newspaper: 76 percent of the seniors in Southport High School say they plan to attend college next year.

STUDY LINK 8·7 Areas of Triangles

Find the area of each triangle.

1.

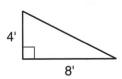

4'

8'

Number model: _____

Area = _____ square feet

2.

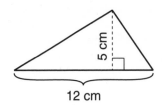

5 cm

12 cm

Number model: _____

Area = _____ square cm

3.

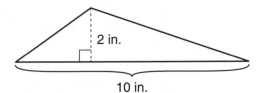

2 in.

10 in.

Number model: _____

Area = _____ square in.

4.

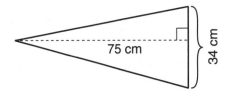

75 cm

34 cm

Number model: _____

Area = _____ square cm

Try This

The area of each triangle is given. Find the length of the base.

5.

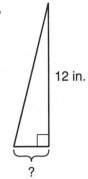

12 in.

?

Area = 18 in²

base = _____ in.

6.

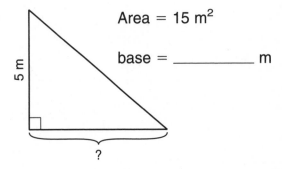

5 m

?

Area = 15 m²

base = _____ m

Practice

7. 18, _____, _____, 45, _____, 63, _____

8. _____, 16, _____, 32, _____, _____, 56

187

STUDY LINK
8·8

Turtle Weights

Turtle	Weight (pounds)
Pacific leatherback	1,552
Atlantic leatherback	1,018
Green sea	783
Loggerhead	568
Alligator snapping	220
Flatback sea	171
Hawksbill sea	138
Kemps Ridley	133
Olive Ridley	110
Common snapping	85

Source: The Top 10 of Everything 2004

1. The Atlantic leatherback is about 10 times heavier than the _____ turtle.

2. The loggerhead is about _____ times the weight of the common snapping turtle.

3. Which turtle weighs about
3 times as much as the loggerhead? _____

4. The flatback sea turtle and the alligator snapping
turtle together weigh about half as much as the _____ turtle.

5. About how many common snapping turtles would
equal the weight of two alligator snapping turtles? _____

6. The Atlantic leatherback is about $\dfrac{\boxed{}}{\boxed{}}$ the weight of the Pacific leatherback.

Practice

Name the factors.

7. 50 _____ **8.** 63 _____

9. 90 _____

Unit 9: Family Letter

Fractions, Decimals, and Percents

In Unit 9, we will be studying percents and their uses in everyday situations. Your child should begin finding examples of percents in newspapers and magazines, on food packages, on clothing labels, and so on, and bring them to class. They will be used to illustrate a variety of percent applications.

As we study percents, your child will learn equivalent values for percents, fractions, and decimals. For example, 50% is equivalent to the fraction $\frac{1}{2}$ and to the decimal 0.5. The class will develop the understanding that **percent** always refers to a **part out of 100.**

Converting "easy" fractions, such as $\frac{1}{2}$, $\frac{1}{5}$, $\frac{1}{10}$, and $\frac{3}{4}$, to decimal and percent equivalents should become automatic for your child. Such fractions are common in percent situations and are helpful with more difficult fractions, decimals, and percents. To help memorize the "easy" fraction/percent equivalencies, your child will play *Fraction/Percent Concentration*.

"Easy" Fractions	Decimals	Percents
$\frac{1}{2}$	0.50	50%
$\frac{1}{4}$	0.25	25%
$\frac{3}{4}$	0.75	75%
$\frac{2}{5}$	0.40	40%
$\frac{7}{10}$	0.70	70%
$\frac{2}{2}$	1.00	100%

Throughout the unit, your child will use a calculator to convert fractions to percents and will learn how to use the percent key $\boxed{\%}$ to calculate discounts, sale prices, and percents of discount.

As part of the World Tour, your child will explore population data, such as literacy rates and percents of people who live in rural and urban areas.

Finally, the class will begin to apply the multiplication and division algorithms to problems that contain decimals. The approach used in *Everyday Mathematics* is straightforward: Students solve the problems as if the numbers were whole numbers. Then they estimate the answers to help them locate the decimal point in the exact answer. In this unit, we begin with fairly simple problems. Your child will solve more difficult problems in *Fifth* and *Sixth Grade Everyday Mathematics*.

Please keep this Family Letter for reference as your child works through Unit 9.

Vocabulary

Important terms in Unit 9:

discount The amount by which the regular price of an item is reduced in a sale, usually given as a fraction or percent of the original price, or as a "percent off."

illiterate An illiterate person cannot read or write.

life expectancy The average number of years a person may be expected to live.

literate A literate person can read and write.

100% box The entire object, the entire collection of objects, or the entire quantity being considered.

```
┌─────────────┐
│ 100% box    │
├─────────────┤
│ 24 books    │
└─────────────┘
```

percent (%) Per hundred or out of a hundred. For example, "48% of the students in the school are boys" means that, on average, 48 out of 100 students in the school are boys; $48\% = \frac{48}{100} = 0.48$

percent of literacy The percent of the total population that is literate; the number of people out of 100 who are able to read and write. For example, 92% of the population in Mexico is literate—this means that, on average, 92 out of 100 people can read and write.

percent or fraction discount The percent or fraction of the regular price that you save in a "percent off" sale. See example under *regular price*.

rank To put in order by size; to sort from smallest to largest or vice versa.

Countries Ranked from Smallest to Largest Percent of Population, Rural		
1	Australia	8%
2	Japan	21%
3	Russia	27%
4	Iran	33%
5	Turkey	34%
6	China	61%
7	Thailand	68%
8	India	72%
9	Vietnam	74%
10	Bangladesh	76%

regular price or list price The price of an item without a discount.

Regular Price	Sale!	Sale Price	You Saved
$19.95	25% OFF	$14.96	$4.99

rural In the country

sale price The amount you pay after subtracting the discount from the regular price. See example under *regular price*.

urban In the city

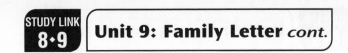

Do-Anytime Activities

To work with your child on the concepts taught in this unit, try these interesting and rewarding activities:

1. Help your child compile a percent portfolio that includes examples of the many ways percents are used in everyday life.

2. Encourage your child to incorporate such terms as "whole," "halves," "thirds," and "fourths" into his or her everyday vocabulary.

3. Practice renaming fractions as percents, and vice versa, in everyday situations. For example, when preparing a meal, quiz your child on what percent $\frac{3}{4}$ of a cup would be.

4. Look through advertisements of sales and discounts. If the original price of an item and the percent of discount are given, have your child calculate the amount of discount and the sale price. If the original price and sale price are given, have your child calculate the amount and percent of discount.

Building Skills through Games

In this unit, your child will play the following games:

Fraction Match See *Student Reference Book,* page 243.
This game is for 2 to 4 players and requires one deck of *Fraction Match* cards. The game develops skill in naming equivalent fractions.

Fraction/Percent Concentration See *Student Reference Book,* page 246.
Two or three players need 1 set of Fraction/Percent Tiles and a calculator to play this game. Playing *Fraction/Percent Concentration* helps students recognize fractions and percents that are equivalent.

Over and Up Squares See *Student Reference Book,* page 257.
This is a game for 2 players and will require a playing grid. The game helps students practice using ordered pairs of numbers to locate points on a rectangular grid.

Polygon Pair-Up See *Student Reference Book,* page 258.
This game provides practice in identifying properties of polygons. It requires a *Polygon Pair-Up* Property Deck and Polygon Deck.

Rugs and Fences See *Student Reference Book,* pages 260 and 261.
This is a game for 2 players and requires a *Rugs and Fences* Polygon Deck, Area and Perimeter Deck, and Record Sheet. The game helps students practice computing the area and perimeter of polygons.

As You Help Your Child with Homework

As your child brings assignments home, you may want to go over the instructions together, clarifying them as necessary. The answers listed below will guide you through this unit's Study Links.

Study Link 9·1

1. $\frac{90}{100}$; 90% **2.** $\frac{53}{100}$; 53% **3.** $\frac{4}{100}$; 4%

4. $\frac{60}{100}$; 0.60 **5.** $\frac{25}{100}$; 0.25 **6.** $\frac{7}{100}$; 0.07

7. 0.50; 50% **8.** 0.75; 75% **9.** 0.06; 6%

Study Link 9·2

1. 100; $\frac{1}{100}$; 0.01; 1% **2.** 20; $\frac{1}{20}$; 0.05; 5%

3. 10; $\frac{1}{10}$; 0.10; 10% **4.** 4; $\frac{1}{4}$; 0.25; 25%

5. 2; $\frac{1}{2}$; 0.50; 50% **6.** 0.75; 75%

7. 0.20; 20%

Study Link 9·3

1.

$\frac{1}{2}$	0	.	5					
$\frac{1}{3}$	0	.	3	3	3	3	3	3
$\frac{1}{4}$	0	.	2	5				
$\frac{1}{5}$	0	.	2					
$\frac{1}{6}$	0	.	1	6	6	6	6	6
$\frac{1}{7}$	0	.	1	4	2	8	5	7
$\frac{1}{8}$	0	.	1	2	5			
$\frac{1}{9}$	0	.	1	1	1	1	1	1
$\frac{1}{10}$	0	.	1					
$\frac{1}{11}$	0	.	0	9	0	9	0	9
$\frac{1}{12}$	0	.	0	8	3	3	3	3
$\frac{1}{13}$	0	.	0	7	6	9	2	3
$\frac{1}{14}$	0	.	0	7	1	4	2	8
$\frac{1}{15}$	0	.	0	6	6	6	6	6
$\frac{1}{16}$	0	.	0	6	2	5		
$\frac{1}{17}$	0	.	0	5	8	8	2	3
$\frac{1}{18}$	0	.	0	5	5	5	5	5
$\frac{1}{19}$	0	.	0	5	2	6	3	1
$\frac{1}{20}$	0	.	0	5				
$\frac{1}{21}$	0	.	0	4	7	6	1	9
$\frac{1}{22}$	0	.	0	4	5	4	5	4
$\frac{1}{23}$	0	.	0	4	3	4	7	8
$\frac{1}{24}$	0	.	0	4	1	6	6	6
$\frac{1}{25}$	0	.	0	4				

Study Link 9·4

1. 34% **2.** 67% **3.** 84% **4.** 52%

5. 85% **6.** 20% **7.** 25% **8.** 30%

9. 62.5% **10.** 70% **11.** 15% **12.** 37.5%

13. Sample answer: I divided the numerator by the denominator and then multiplied by 100.

14. 86% **15.** 3% **16.** 14% **17.** 83.5%

Study Link 9·5

1. 7%; 7%; 7%; 8%; 10%; 11%; 10%; 10%; 9%; 8%; 7%

3. Sample answer: I divided the number of marriages for each month by the total number of marriages, then multiplied by 100 and rounded to the nearest whole number.

Study Link 9·6

1. The varsity team. They won $\frac{8}{10}$ or 80% of their games. The junior varsity team only won $\frac{6}{8}$ or 75% of their games.

2. 2: 11; $\frac{5}{11}$; 45%   3: 3; $\frac{3}{3}$; 100%

4: 11; $\frac{9}{11}$; 82%   5: 7; $\frac{4}{7}$; 57%

6: 16; $\frac{11}{16}$; 69%   7: 10; $\frac{6}{10}$; 60%

8: 2; $\frac{1}{2}$; 50%

Study Link 9·7

1. 50% **2.** Tuvalu **3.** 5%

4. Dominica; Antigua and Barbuda; and Palau

5. 300%

Study Link 9·8

1. 25.8 **2.** 489.6 **3.** 45.12 **4.** 112.64

7. Sample answer: I estimated that the answer should be about 5 * 20 = 100.

8. 212.4 **9.** 38.64 **10.** 382.13

Study Link 9·9

1. 14.8 **2.** 0.2700 **3.** 24.96 **4.** 0.860

5. 23.4 **6.** 58.32

7. Sample answer: I estimated that the answer should be about $\frac{100}{4}$ = 25.

8. 4.2 **9.** 38.7 **10.** 0.65

STUDY LINK 9·1 Fractions, Decimals, and Percents

SRB
61 62

Rename each decimal as a fraction and a percent.

1. $0.90 = \dfrac{\square}{100} = $ _____ %   **2.** $0.53 = \dfrac{\square}{100} = $ _____ %   **3.** $0.04 = \dfrac{\square}{100} = $ _____ %

Rename each percent as a fraction and a decimal.

4. $60\% = \dfrac{\square}{100} = $ __._____   **5.** $25\% = \dfrac{\square}{100} = $ __._____   **6.** $7\% = \dfrac{\square}{100} = $ __._____

Rename each fraction as a decimal and a percent.

7. $\dfrac{50}{100} = $ __._____ $= $ _____ %   **8.** $\dfrac{75}{100} = $ __._____ $= $ _____ %   **9.** $\dfrac{6}{100} = $ __._____ $= $ __ %

10. Shade more than $\dfrac{10}{100}$ and less than $\dfrac{30}{100}$ of the grid.
Write the value of the shaded part as a decimal and a percent.

Decimal: _____

Percent: _____

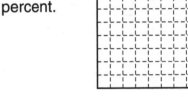

11. Shade more than 25% and less than 60% of the grid.
Write the value of the shaded part as a decimal and a percent.

Decimal: _____

Percent: _____

12. Shade more than 0.65 and less than 0.85 of the grid.
Write the value of the shaded part as a decimal and a percent.

Decimal: _____

Percent: _____

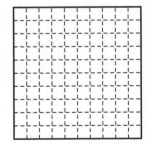

Practice

Order the fractions from smallest to largest.

13. $\dfrac{3}{6}, \dfrac{3}{3}, \dfrac{3}{5}, \dfrac{3}{7}$ _____   **14.** $\dfrac{2}{3}, \dfrac{6}{7}, \dfrac{1}{2}, \dfrac{19}{20}$ _____

STUDY LINK 9·1   Trivia Survey

Conduct the survey below. The results will be used in Lesson 9-6.

Find at least five people to answer the following survey questions. You can ask family members, relatives, neighbors, and friends.

BE CAREFUL! You will not ask every person every question. Pay attention to the instructions that go with each question.

Record each answer with a tally mark in the Yes or No column.

Question	Yes	No
1. Is Monday your favorite day? (Ask everyone younger than 20.)		
2. Have you gone to the movies in the last month? (Ask everyone older than 8.)		
3. Did you eat breakfast today? (Ask everyone over 25.)		
4. Do you keep a map in your car? (Ask everyone who owns a car.)		
5. Did you eat at a fast-food restaurant yesterday? (Ask everyone.)		
6. Did you read a book during the last month? (Ask everyone over 20.)		
7. Are you more than 1 meter tall? (Ask everyone over 20.)		
8. Do you like liver? (Ask everyone.)		

STUDY LINK 9·2 | **Coins as Percents of $1**

1. How many pennies in $1? _____ What fraction of $1 is 1 penny? _____

   Write the decimal that shows what part of $1 is 1 penny. _____

   What percent of $1 is 1 penny? _____%

2. How many nickels in $1? _____ What fraction of $1 is 1 nickel? _____

   Write the decimal that shows what part of $1 is 1 nickel. _____

   What percent of $1 is 1 nickel? _____%

3. How many dimes in $1? _____ What fraction of $1 is 1 dime? _____

   Write the decimal that shows what part of $1 is 1 dime. _____

   What percent of $1 is 1 dime? _____%

4. How many quarters in $1? _____ What fraction of $1 is 1 quarter? _____

   Write the decimal that shows what part of $1 is 1 quarter. _____

   What percent of $1 is 1 quarter? _____%

5. How many half-dollars in $1? _____ What fraction of $1 is 1 half-dollar? _____

   Write the decimal that shows what part of $1 is 1 half-dollar. _____

   What percent of $1 is 1 half-dollar? _____%

6. Three quarters (75¢) is $\frac{3}{4}$ of $1.

   Write the decimal. _____

   What percent of $1 is

   3 quarters? _____%

7. Two dimes (20¢) is $\frac{2}{10}$ of $1.

   Write the decimal. _____

   What percent of $1 is

   2 dimes? _____%

Practice

8. _____ = 748 * 6   9. 51 * 90 = _____   10. _____ = 28 * 903

STUDY LINK 9·3 **Calculator Decimals**

1. Use your calculator to rename each fraction below as a decimal.

$\frac{1}{2}$	0	.	5					
$\frac{1}{3}$	0	.	3	3	3	3	3	3
$\frac{1}{4}$								
$\frac{1}{5}$								
$\frac{1}{6}$								
$\frac{1}{7}$								
$\frac{1}{8}$								
$\frac{1}{9}$								
$\frac{1}{10}$								
$\frac{1}{11}$								
$\frac{1}{12}$								
$\frac{1}{13}$								

$\frac{1}{14}$								
$\frac{1}{15}$								
$\frac{1}{16}$								
$\frac{1}{17}$								
$\frac{1}{18}$								
$\frac{1}{19}$								
$\frac{1}{20}$								
$\frac{1}{21}$								
$\frac{1}{22}$								
$\frac{1}{23}$								
$\frac{1}{24}$								
$\frac{1}{25}$								

2. Make up some of your own.

$\frac{1}{73}$	0	.	0	1	3	6	9	8
$\frac{1}{}$								
$\frac{1}{}$								

$\frac{1}{}$							
$\frac{1}{}$							
$\frac{1}{}$							

Practice

3. $6\overline{)96}$ = _____

4. 91 / 5 = _____

5. _____ = 864 ÷ 8

6. 575 ÷ 7 = _____

199

STUDY LINK 9·4 Fractions and Decimals to Percents

Do NOT use a calculator to convert these fractions to percents.
On the back of this page, show your work for Problems 3–6.

1. $\dfrac{34}{100} =$ _____%

2. $\dfrac{67}{100} =$ _____%

3. $\dfrac{42}{50} =$ _____%

4. $\dfrac{13}{25} =$ _____%

5. $\dfrac{17}{20} =$ _____%

6. $\dfrac{25}{125} =$ _____%

Use a calculator to convert these fractions to percents.

7. $\dfrac{23}{92} =$ _____%

8. $\dfrac{12}{40} =$ _____%

9. $\dfrac{20}{32} =$ _____%

10. $\dfrac{49}{70} =$ _____%

11. $\dfrac{60}{400} =$ _____%

12. $\dfrac{21}{56} =$ _____%

13. Describe how you used your calculator to convert the fractions
in Problems 7–12 to percents.

_____

_____

_____

Do NOT use a calculator to convert these decimals to percents.

14. 0.86 = _____%

15. 0.03 = _____%

16. 0.140 = _____%

17. 0.835 = _____%

Practice

Order the fractions from smallest to largest.

18. $\dfrac{7}{16}, \dfrac{7}{8}, \dfrac{7}{12}, \dfrac{7}{9}$ _____

19. $\dfrac{7}{15}, \dfrac{3}{15}, \dfrac{8}{15}, \dfrac{4}{15}$ _____

20. $\dfrac{5}{9}, \dfrac{15}{16}, \dfrac{1}{4}, \dfrac{9}{10}$ _____

STUDY LINK 9·5 | **Renaming Fractions as Percents**

In 2001, there were about 2,317,000 marriages in the United States.
The table below shows the approximate number of marriages each month.

1. Use a calculator to find the percent of the total number of marriages that
   occurred each month. Round the answers to the nearest whole-number percent.

Month	Approximate Number of Marriages	Approximate Percent of Total Marriages
January	147,000	6%
February	159,000	
March	166,000	
April	166,000	
May	189,000	
June	237,000	
July	244,000	
August	225,000	
September	224,000	
October	217,000	
November	191,000	
December	152,000	

Source: U.S. Department of Health and Human Services

2. According to the table, what is the most popular month for a wedding? _____

   What is the least popular month for a wedding? _____

3. Describe how you used your calculator to find the percent for each month.

   _____

   _____

Practice

Name all the factors of each number.

4. 63 _____

5. 28 _____

STUDY LINK 9·6 Use Percents to Compare Fractions

SRB
62 207

1. The girls' varsity basketball team won 8 of the 10 games it played. The junior varsity team won 6 of 8 games. Which team has the better record? Explain your reasoning.

_____

_____

2. Complete the table of shots taken (not including free throws) during a game. Calculate the percent of shots made to the nearest whole percent.

Player	Shots Made	Shots Missed	Total Shots	Shots Made / Total Shots	% of Shots Made
1	5	12	17	$\frac{5}{17}$	29%
2	5	6			
3	3	0			
4	9	2			
5	4	3			
6	11	5			
7	6	4			
8	1	1			

3. The basketball game is tied. Your team has the ball. There is only enough time for one more shot. Based only on the information in the table, which player would you choose to take the shot? Why?

_____

_____

_____

_____

Practice

4. $\frac{1}{3} + \frac{1}{6} =$ _____

5. _____ $= \frac{3}{4} - \frac{1}{2}$

6. _____ $= \frac{7}{10} + \frac{1}{5}$

7. $\frac{5}{8} - \frac{1}{4} =$ _____

STUDY LINK 9·7 Least-Populated Countries

The table below shows the approximate population for the 10 least-populated countries in the world. Use the data to estimate answers to the problems.

Country	Population
Vatican City	900
Tuvalu	11,000
Nauru	13,000
Palau	20,000
San Marino	28,000
Monaco	32,000
Liechtenstein	33,000
St. Kitts and Nevis	39,000
Antigua and Barbuda	68,000
Dominica	69,000

Source: Top Ten of Everything 2004

1. The population of Liechtenstein is about _____% of the population of Dominica.

2. What country's population is about 33% of Liechtenstein's population? _____

3. The population of Vatican City is about _____% of the population of Palau.

4. The population of the 10 countries listed is 314,900. What 3 country populations together equal about 50% of that total?

_____

5. The population of St. Kitts and Nevis is about _____% of Nauru's population.

Practice

6. 27 * 4 = _____

7. _____ = 508 * 8

8. _____ = 63 * 86

9. 849 * 52 = _____

STUDY LINK 9·8 | Multiplying Decimals

For each problem below, the multiplication has been done correctly, but the decimal point is missing in the answer. Correctly place the decimal point in the answer.

1. 6 * 4.3 = 2 5 8

2. 72 * 6.8 = 4 8 9 6

3. 0.96 * 47 = 4 5 1 2

4. 5.12 * 22 = 1 1 2 6 4

5. 8,457 * 9.8 = 8 2 8 7 8 6

6. 0.04 * 140 = 5 6

7. Explain how you decided where to place the decimal point in Problem 4.

_____

_____

_____

Try This

Multiply. Show your work.

8. 5.9 * 36 = _____	**9.** 0.46 * 84 = _____	**10.** _____ = 7.21 * 53

Practice

11. _____ = 96 ÷ 6

12. 4)̄67 = _____

13. _____ = 411 / 3

14. 9)̄903 = _____

STUDY LINK 9·9 | Dividing Decimals

For each problem below, the division has been done correctly,
but the decimal point is missing in the answer. Correctly place the
decimal point in the answer.

1. 88.8 / 6 = 1 4 8

2. 1.35 / 5 = 2 7 0 0

3. 99.84 / 4 = 2 4 9 6

4. 2.58 / 3 = 8 6 0

5. 163.8 / 7 = 2 3 4

6. 233.28 / 4 = 5 8 3 2

7. Explain how you decided where to place the decimal point in Problem 3.

_____

_____

_____

Try This

Divide. Show your work.

8. 6)25.2	**9.** 4)154.8	**10.** 9)5.85
Answer: _____	Answer: _____	Answer: _____

Practice

11. _____ = $\frac{5}{8} + \frac{2}{8}$ **12.** $\frac{5}{9} - \frac{1}{3}$ = _____   **13.** _____ = $\frac{7}{10} + \frac{2}{10}$ **14.** $\frac{9}{10} - \frac{1}{2}$ = _____

STUDY LINK 9·10

Unit 10: Family Letter

Reflections and Symmetry

In this unit, your child will take another look at geometry, with an emphasis on symmetry. Many objects in nature are symmetrical: flowers, insects, and the human body, to name just a few. Symmetry is all around—in buildings, furniture, clothing, and paintings.

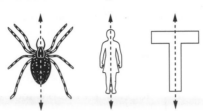

The class will focus on **reflectional symmetry**, also called **line symmetry** or **mirror symmetry**, in which half of a figure is the mirror image of the other half. Encourage your child to look for symmetrical objects, and if possible, to collect pictures of symmetrical objects from magazines and newspapers. For example, the right half of the printed letter T is the mirror image of the left half. If you have a small hand mirror, have your child check letters, numbers, and other objects to see whether they have line symmetry. The class will use a device called a **transparent mirror,** which is pictured below. Students will use it to see and trace the mirror image of an object.

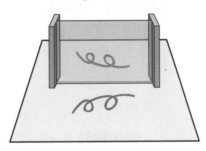

Geometry is not only the study of figures (such as lines, rectangles, and circles), but also the study of transformations or "motions" of figures. These motions include **reflections** (flips), **rotations** (turns), and **translations** (slides). Your child will use these motions to create pictures like the ones below, called **frieze patterns.**

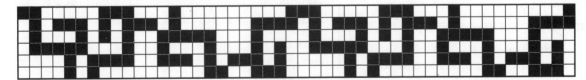

Students will also work with positive and negative numbers, looking at them as reflections of each other across zero on a number line. They will develop skills of adding positive and negative numbers by thinking in terms of credits and debits for a new company, and they will practice these skills in the *Credits/Debits Game.*

Please keep this Family Letter for reference as your child works through Unit 10.

Vocabulary

Important terms in Unit 10:

frieze pattern A geometric design in a long strip in which an element is repeated over and over. The element may be rotated, translated, and reflected. Frieze patterns are often found on the walls of buildings, on the borders of rugs and tiled floors, and on clothing.

image The reflection of an object that you see when you look in the mirror. Also a figure that is produced by a transformation (reflection, translation, or rotation) of another figure. See *preimage*.

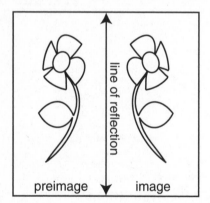

preimage   image

line of reflection A line halfway between a figure (preimage) and its reflected image. In a reflection, a figure is "flipped over" the line of reflection.

line of symmetry A line drawn through a figure that divides the figure into two parts that are mirror images of each other. The two parts look alike, but face in opposite directions.

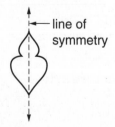

line of symmetry

negative number A number that is less than zero; a number to the left of zero on a horizontal

number line or below zero on a vertical number line. The symbol "−" may be used to write a negative number. For example, "negative 5" is usually written as −5.

preimage A geometric figure that is somehow changed (by a *reflection*, a *rotation*, or a *translation*, for example) to produce another figure. See *image*.

reflection (flip) The "flipping" of a figure over a line (the *line of reflection*) so that its image is the mirror image of the original (preimage).

reflection

rotation (turn) A movement of a figure around a fixed point, or axis; a "turn."

symmetric Having the same size and shape on either side of a line, or looking the same when turned by some amount less than 360°.

transformation Something done to a geometric figure that produces a new figure. The most common transformations are translations (slides), reflections (flips), and rotations (turns).

translation A movement of a figure along a straight line; a "slide." In a translation, each point of the figure slides the same distance in the same direction.

translation

Do-Anytime Activities

To work with your child on concepts taught in this unit, try these interesting and rewarding activities:

1. Have your child look for frieze patterns on buildings, rugs, floors, and clothing. If possible, have your child bring pictures to school or make sketches of friezes that he or she sees.

2. Encourage your child to study the mathematical qualities of the patterns of musical notes and rhythms. Composers of even the simplest of tunes use reflections and translations of notes and chords (groups of notes).

3. Encourage your child to incorporate transformation vocabulary—**symmetric, reflected, rotated,** and **translated**—into his or her everyday vocabulary.

Building Skills through Games

In this unit, your child will play the following games to develop his or her understanding of addition and subtraction of positive and negative numbers, practice estimating and measuring angles, practice plotting ordered pairs in the first quadrant of a coordinate grid, and identify properties of polygons. For detailed instructions, see the *Student Reference Book.*

Angle Tangle See *Student Reference Book,* page 230. Two players need a protactor, straightedge, and several sheets of blank paper to play this game. This game provides practice estimating and measuring angle sizes.

Credits/Debits Game See *Student Reference Book,* page 238. Playing the *Credits/Debits Game* offers students practice adding and subtracting positive and negative numbers.

Over and Up Squares See *Student Reference Book,* page 257. Two players need a gameboard and record sheet, 2 different-colored pencils, and 2 six-sided dice to play this game. Playing this game provides practice plotting ordered pairs and developing a winning game strategy.

Polygon Pair-Up See *Student Reference Book,* page 258. To play this game, two players need a deck of polygon cards, a deck of property cards, and paper and pencils for sketching. Playing this game provides students with practice identifying properties of polygons.

As You Help Your Child with Homework

As your child brings assignments home, you may want to go over the instructions together, clarifying them as necessary. The answers listed below will guide you through some of the Study Links in this unit.

Study Link 10·2

1.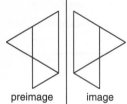

preimage   image

3. preimage
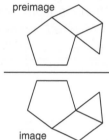
image

5.
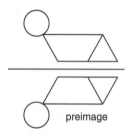
preimage

Study Link 10·3

1. preimage   image
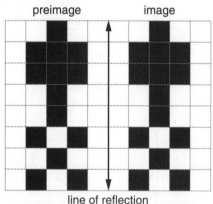
line of reflection

3. preimage   image
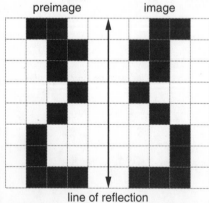
line of reflection

Study Link 10·4

2.
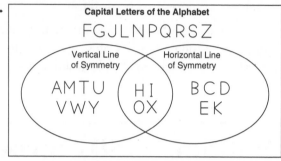

Capital Letters of the Alphabet
FGJLNPQRSZ

Vertical Line of Symmetry   Horizontal Line of Symmetry

AMTU VWY   HI OX   BCD EK

3. Sample answers:

horizontal	vertical
BOX	TAX
KID	YOU
BOOK	MAT
KICK	HIM

Study Link 10·5

1. a. reflection   **b.** translation   **c.** rotation

Study Link 10·6

1. <   **2.** <   **3.** <   **4.** >

5. $-8, -3.4, -\frac{1}{4}, \frac{1}{2}, 1.7, 5$

6. $-43, -3, 0, \frac{14}{7}, 5, 22$

7. Sample answers: $\frac{1}{4}, \frac{1}{2}, \frac{3}{4}, 1$

8. Sample answers: $-2, -1, -\frac{1}{2}, -\frac{1}{4}$

9. a. 13   **b.** -5 **c.** -13

10. a. 8   **b.** -2 **c.** -8

11. a. 15   **b.** 11   **c.** -15

STUDY LINK
10·1
A Reflected Image

There is a simple design in the box in the middle of this page. It is the **preimage.**

Hold this page in front of a mirror, with the printed side facing the mirror. On a blank piece of paper, sketch what the design looks like in the mirror—the **image.**

Compare your sketch (image) with the design on the Study Link page (preimage). Bring both the preimage and image to school tomorrow.

SRB
106

Sketch the design as it looks in the mirror.

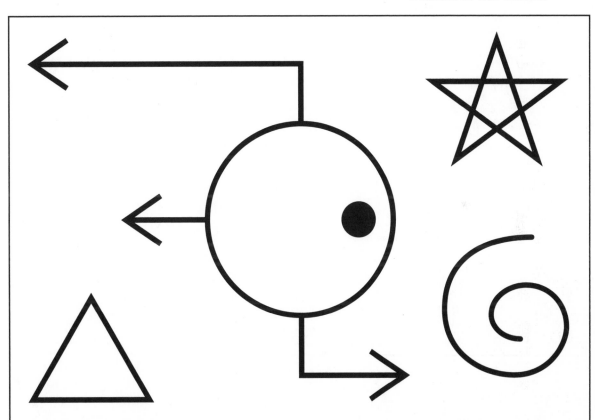

Practice

1. 10% of 130 = _____

2. _____ = 25% of 32

3. _____ = 15% of 120

4. 70% of 490 = _____

STUDY LINK
10·2

Lines of Reflection

For each preimage and image, draw the line of reflection.

1.

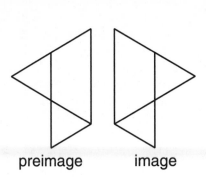

preimage   image

2.

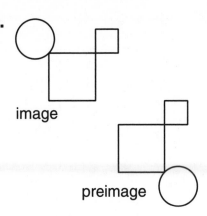

image

preimage

3.

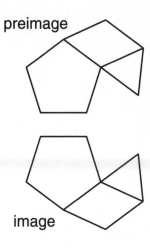

preimage

image

For each preimage, use your Geometry Template to draw the image on the other side of the line of reflection.

4.

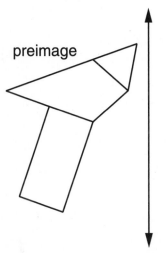

preimage

5.

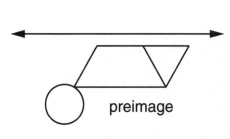

preimage

6.

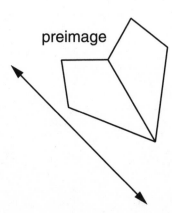

preimage

7. Create one of your own.
preimage

219

STUDY LINK
10·3 **Reflections**

Shade squares to create the reflected image of each preimage.

1. preimage   image

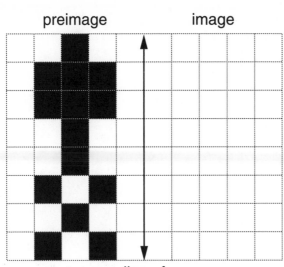

line of
reflection

2. image   preimage

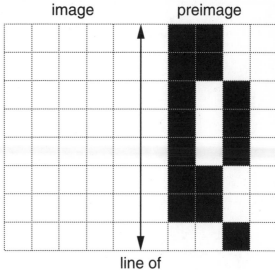

line of
reflection

3. preimage   image

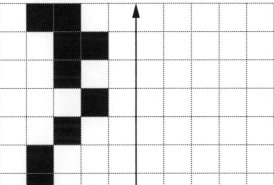

line of
reflection

4. image   preimage

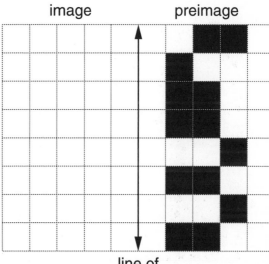

line of
reflection

Practice

5. 54 * 6 = _____

6. 29 * 36 = _____

7. _____ = 45 * 45

8. _____ = 837 * 63

221

STUDY LINK 10·4 | Line Symmetry in the Alphabet

SRB
109

1. Print the 26 capital letters of the alphabet below.

— — — — — — — — — — — — —

— — — — — — — — — — — — —

2. The capital letter A has a vertical line of symmetry.

The capital letter B has a horizontal line of symmetry.

Use the letters of the alphabet to complete the Venn diagram.

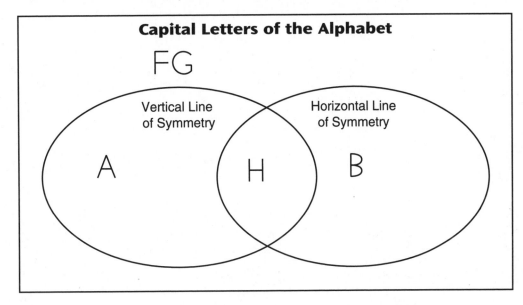

Capital Letters of the Alphabet

FG

Vertical Line of Symmetry

Horizontal Line of Symmetry

A   H   B

3. The word BED has a horizontal line of symmetry.

The word HIT has a vertical line of symmetry.

Use capital letters to list words that have horizontal or vertical line symmetry.

horizontal   **vertical**

_____ _____   _____ _____

_____ _____   _____ _____

Practice

4. $86 \div 9 =$ _____

5. _____ $= 68 / 4$

6. $6\overline{)742} =$ _____

7. _____ $= 855 / 7$

223

STUDY LINK
10·5

Geometric Patterns

SRB
106–108

1. Continue each pattern. Then tell if you continued the pattern by using a reflection, rotation, or translation of the original design.

a. _____

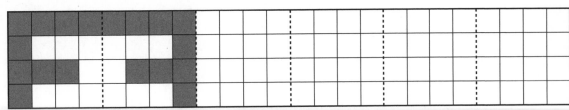

b. _____

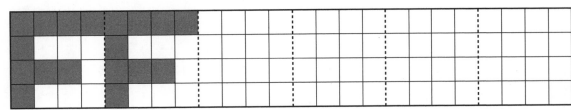

c. _____

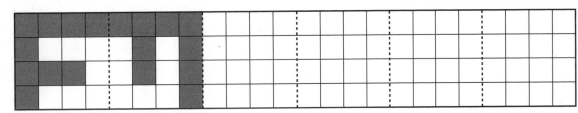

2. Make up your own pattern.

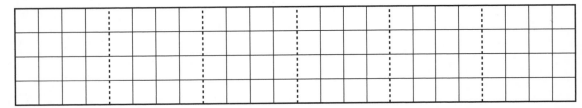

Practice

3. 50% of $25.00 = _____

4. 25% of $10.00 = _____

5. _____ = 40% of $150.00

6. _____ = 20% of $250.00

STUDY LINK 10·6 Positive and Negative Numbers

Write < or > to make a true number sentence.

1. 3 _____ 14   **2.** −7 _____ 7   **3.** 19 _____ 20   **4.** −8 _____ −10

List the numbers in order from least to greatest.

5. 5, −8, $\frac{1}{2}$, −$\frac{1}{4}$, 1.7, −3.4

_____ _____ _____ _____ _____ _____

least   **greatest**

6. −43, 22, $\frac{14}{7}$, 5, −3, 0

_____ _____ _____ _____ _____ _____

least   **greatest**

7. Name four positive numbers less than 2.   _____ _____ _____ _____

8. Name four negative numbers greater than −3.   _____ _____ _____ _____

Use the number line to help you solve Problems 9–11.

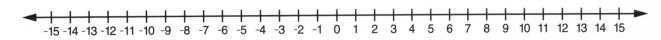

9. a. 4 + 9 = _____   **b.** 4 + (−9) = _____   **c.** (−4) + (−9) = _____

10. a. 5 + 3 = _____   **b.** (−5) + 3 = _____   **c.** (−5) + (−3) = _____

11. a. _____ = 2 + 13   **b.** _____ = (−2) + 13   **c.** _____ = (−2) + (−13)

Practice

12. 1.02 + 12.88 = _____   **13.** 7.26 − 1.94 = _____

14. _____ + 5.84 = 8.75   **15.** 3.38 − _____ = 2.62

227

STUDY LINK
10·7

Unit 11: Family Letter

3-D Shapes, Weight, Volume, and Capacity

Our next unit introduces several new topics, as well as reviewing some of the work with geometric solids from previous grades and some of the main ideas your child has been studying this past year.

We begin with a lesson on weight, focusing on grams and ounces. Students handle and weigh a variety of objects, trying to develop "weight sense" so that they can estimate weights effectively. The class participates in creating a Gram & Ounce Museum by displaying everyday objects labeled with their weights.

As part of a review of the properties of 3-dimensional shapes (prisms, pyramids, cylinders, and cones), your child will construct models of geometric solids using straws and paper patterns. They will use these models as they discuss vocabulary such as *face, edge,* and *vertex* and compare features of geometric solids.

By experimenting with cubes, the class will develop and apply a formula for finding the volumes of rectangular prisms (solids that look like boxes).

We will consider familiar units of capacity (cups, pints, quarts, gallons) and the relationships among them.

Your child will also explore subtraction of positive and negative numbers by playing a variation of the *Credits/Debits Game* introduced in Unit 10.

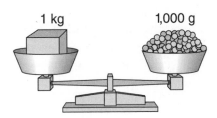

1 kg   1,000 g

In Lesson 11-1, a pan balance is used to measure weight in grams.

Please keep this Family Letter for reference as your child works through Unit 11.

Vocabulary

Important terms in Unit 11:

capacity (1) The amount of space occupied by a 3-dimensional shape. Same as *volume*. (2) Less formally, the amount a container can hold. Capacity is often measured in units such as quarts, gallons, cups, or liters. (3) The maximum *weight* a scale can measure.

cone A 3-dimensional shape that has a circular base, a *curved surface,* and one vertex, which is called the apex. The points on the curved surface of a cone are on straight lines connecting the apex and the circumference of the base.

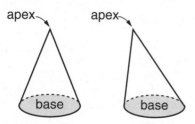

cubic unit A unit used in measuring *volume,* such as a cubic centimeter or a cubic foot.

curved surface A 2-dimensional surface that is rounded rather than flat. Spheres, *cylinders,* and *cones* each have one curved surface.

cylinder A 3-dimensional shape that has two circular or elliptical bases that are parallel and congruent and are connected by a *curved surface.* A can is shaped like a cylinder.

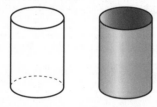

dimension A measure along one direction of an object, typically length, width, or height. For example, the dimensions of a box might be 24 cm by 20 cm by 10 cm.

formula A general rule for finding the value of something. A formula is often written using letters, called variables, that stand for the quantities involved.

geometric solid The surface or surfaces that make up a 3-dimensional shape, such as a *prism, cylinder, cone,* or sphere. Despite its name, a geometric solid is hollow; it does not contain the points in its interior.

prism A 3-dimensional shape with two parallel and congruent polygonal regions for bases and lateral faces formed by all the line segments with endpoints on corresponding edges of the bases. The lateral faces are all parallelograms.

triangular prism   rectangular prism   hexagonal prism

pyramid A 3-dimensional shape with a polygonal region for a base, a point (apex) not in the plane of the base, and all of the line segments with one endpoint at the apex and the other on an edge of the base. All faces except the base are triangular.

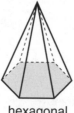

hexagonal pyramid   rectangular pyramid

3-dimensional (3-D) shape A shape whose points are not all in a single plane. Examples include *prisms, pyramids,* and spheres, all of which have length, width, and height.

volume The amount of space occupied by a 3-dimensional shape. Same as *capacity.* The amount a container can hold. Volume is often measured in cubic units, such as cm³, cubic inches, or cubic feet.

weight A measure of the force of gravity on an object. Weight is measured in metric units such as grams, kilograms, and milligrams and in U.S. customary units such as pounds and ounces.

Do-Anytime Activities

To work with your child on the concepts taught in this unit, try these interesting and rewarding activities:

1. Have your child compile a list of the world's heaviest objects or things. For example, which animal has the heaviest baby? What is the world's heaviest human-made structure? What is the greatest amount of weight ever hoisted by a person?

2. Have your child compile a portfolio of 3-dimensional shapes. Images can be taken from newspapers, magazines, photographs, and so on.

3. Encourage your child to create his or her own mnemonics and/or sayings for converting between units of capacity and weight. One such example is the old English saying "A pint's a pound the world around." (1 pint = 16 oz = 1 lb)

Building Skills through Games

In Unit 11, your child will play the following games. For detailed instructions, see the *Student Reference Book.*

Chances Are See *Student Reference Book,* page 236.
This game is for 2 players and requires one deck of *Chances Are* Event Cards and one deck of *Chances Are* Probability Cards. The game develops skill in using probability terms to describe the likelihood of events.

Credits/Debits Game See *Student Reference Book,* page 238.
This is a game for 2 players. Game materials include 1 complete deck of number cards and a recording sheet. The *Credits/Debits Game* helps students practice addition of positive and negative integers.

Credits/Debits Game (**Advanced Version**) See *Student Reference Book,* page 239.
This game is similar to the *Credits/Debits Game* and helps students practice addition and subtraction of positive and negative integers.

As You Help Your Child with Homework

As your child brings assignments home, you may want to go over the instructions together, clarifying them as necessary. The answers listed below will guide you through this unit's Study Links.

Study Link 11·1

1. 59   2. 96,640

3. Bagel and pumpkin; or taco and gingerbread man

4. Pasta, Chocolate bar, Hamburger, Ice cream sundae

6. −$50   7. −$75   8. $0

9. $30

Study Link 11·2

1. **a.** square pyramid   **b.** cone

   c. sphere   **d.** cylinder

   e. rectangular prism **f.** triangular prism

2.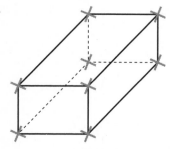

3. 6   4. 7,000; 63,560; and 91

5. 24; 120; 600

Study Link 11·3

1. cone   2. square pyramid

3. hexagonal prism   4. octahedron

6. $10   7. −$70

8. −$15   9. −$100

10. −$55   11. −$400

Study Link 11·4

4. 24   5. 17 R1, or $17\frac{1}{5}$

6. 29   7. 89 R2, or $89\frac{2}{4}$

Study Link 11·5

1. **a.** 39   **b.** 30

2. **a.** (3 ∗ 3) ∗ 6 = 54; 54

   b. (2 ∗ 5) ∗ 9.7 = 97; 97

3. **a.** 150   **b.** 150

4. −49   5. −40   6. 29   7. 73

Study Link 11·6

1. −110   2. −8   3. −8

4. 15   5. 14   6. −19

7. −70   8. 18

11. <   12. <   13. >

14. >   15. >   16. >

17. −14, −2.5, −0.7, $\frac{30}{6}$, 5.6, 8

18. −7, −$\frac{24}{6}$, −$\frac{3}{5}$, 0.02, 0.46, 4

19. 2,652   20. 44,114   21. 158

22. 106 R4, or $106\frac{4}{7}$

Study Link 11·7

Answers vary for Problems 1–4.

5. 4   6. 48   7. 2

8. 3   9. 3   10. 10

11. 4   12. −4   13. −40

14. −120

STUDY LINK 11·1

The World's Largest Foods

Food	Weight	Date	Location
Apple	3 pounds 11 ounces	October 1997	Linton, England
Bagel	714 pounds	July 1998	Mattoon, Illinois
Bowl of pasta	7,355 pounds	February 2004	Hartford, New York
Chocolate bar	5,026 pounds	March 2000	Turin, Italy
Garlic	2 pounds 10 ounces	1985	Eureka, California
Gingerbread man	372.13 pounds	November 2003	Vancouver, Canada
Hamburger	6,040 pounds	September 1999	Sac, Montana
Ice cream sundae	22.59 tons	July 1988	Alberta, Canada
Pumpkin	1,337 pounds	October 2002	Topsfield, Massachusetts
Taco	1,654 pounds	March 2003	Mexicali, Mexico

Source: www.guinnessworldrecords.com

Use the information in the table to solve the following problems.

1. The largest apple weighed _____ ounces.

2. A typical hamburger weighs about 4 ounces. The largest hamburger weighed

_____ ounces.

3. Which 2 foods together weigh about a ton? _____ and

_____

4. A kilogram is a little more than 2 pounds. Which 4 foods each weigh more than 1,000 kilograms?

_____

5. On the back of this page, use data from the table to write and solve your own problem.

| **Practice** |

6. −$75 + $25 = _____

7. _____ = −$45 + (−$30)

8. _____ = −$60 + $60

9. $55 + (−$25) = _____

233

STUDY LINK 11·2 | **Solids**

1. The pictures below show objects that are shaped approximately like geometric solids. Identify each object as one of the following: **cylinder, cone, sphere, triangular prism, square pyramid,** or **rectangular prism.**

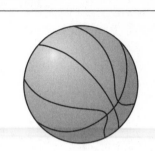

a.	**b.**	**c.**
Type: _____ _____	Type: _____ _____	Type: _____ _____

d.	**e.**	**f.**
Type: _____ _____	Type: _____ _____	Type: _____ _____

  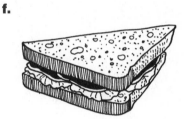

2. Mark *X*s on the vertices of the rectangular prism.

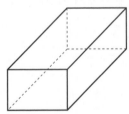

3. How many edges does the tetrahedron have? _____ edges

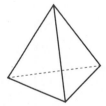

Practice

4. Circle the numbers that are multiples of 7.   132   7,000   63   560   834   91

5. Circle the numbers that are multiples of 12.   24   120   38   600   100   75

235

STUDY LINK 11·3 | **Geometry Riddles**

Answer the following riddles.

1. I am a geometric solid.
I have two surfaces.
One of my surfaces is formed by a circle.
The other surface is curved.

What am I? _____

2. I am a geometric solid.
I have one square base.
I have four triangular faces.
Some Egyptian pharaohs were buried
in tombs shaped like me.

What am I? _____

3. I am a polyhedron.
I am a prism.
My two bases are hexagons.
My other faces are rectangles.

What am I? _____

4. I am a polyhedron.
All of my faces are the same.
All of my faces are equilateral triangles.
I have eight faces.

What am I? _____

Try This

5. Write your own geometry riddle.

_____

_____

_____

_____

_____

_____

_____

Practice

6. $-\$20 + \$30 =$ _____

7. _____ $= -\$35 + (-\$35)$

8. _____ $= \$10 + (-\$25)$

9. $\$0 + (-\$100) =$ _____

10. $-\$15 + (-\$40) =$ _____

11. _____ $= -\$300 + (-\$100)$

237

STUDY LINK 11·4 | Volume

Cut out the pattern below and tape it together to form an open box.

1. Find and record two items in your home that have volumes equal to about $\frac{1}{2}$ of the volume of the open box.

   _____   _____

2. Find and record two items in your home that have about the same volume as the open box.

   _____   _____

3. Find and record two items in your home that have volumes equal to about 2 times the volume of the open box.

   _____   _____

Practice

4. 96 ÷ 4 = _____

5. 86 / 5 = _____

6. $\frac{232}{8}$ = _____

7. 4)‾358‾ = _____

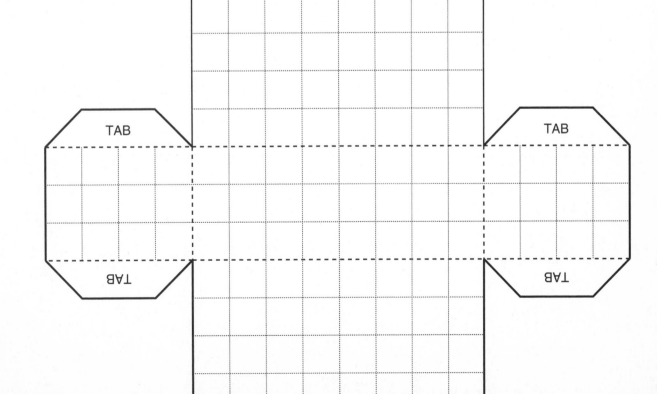

239

STUDY LINK 11·5 | **Volume**

SRB 137 138

1. Find the volume of each stack of centimeter cubes.

a.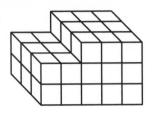

Volume = _____ cm^3

b.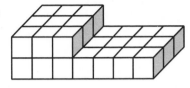

Volume = _____ cm^3

2. Calculate the volume of each rectangular prism.

a.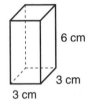

6 cm

3 cm

3 cm

Number model: _____

Volume = _____ cm^3

b.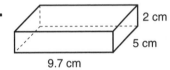

2 cm

5 cm

9.7 cm

Number model: _____

Volume = _____ cm^3

3. What is the total number of cubes needed to completely fill each box?

a.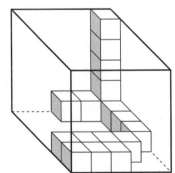

_____ cubes

b.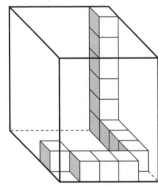

_____ cubes

Practice

4. $-65 + 16 =$ _____

5. _____ $= -21 + (-19)$

6. _____ $= 84 + (-55)$

7. $-16 + 89 =$ _____

241

STUDY LINK
11·6

Positive and Negative Numbers

Add or subtract.

1. $-40 + (-70) =$ _____

2. $12 - 20 =$ _____

3. _____ $= -14 - (-6)$

4. _____ $= 10 - (-5)$

5. $15 + (-1) =$ _____

6. $-12 - 7 =$ _____

7. _____ $= 60 + (-130)$

8. _____ $= -2 - (-20)$

9. Write two subtraction problems with an answer of -8.

   _____ $-$ _____ $= -8$ _____ $-$ _____ $= -8$

10. Write two addition problems with an answer of -30.

   _____ $+$ _____ $= -30$ _____ $+$ _____ $= -30$

Write $<$ or $>$ to make a true number sentence.

11. $0 - 7$ _____ -6

12. -11 _____ $-13 - (-5)$

13. $7 + (-2)$ _____ -8

14. $18 + (-8)$ _____ -18

15. $26 - (-14)$ _____ $27 + (-16)$

16. $9 - (-11)$ _____ $0 + (-20)$

List the numbers in order from least to greatest.

17. $\frac{30}{6}$, 8, -14, -0.7, 5.6, -2.5

   _____ _____ _____ _____ _____ _____
   least   greatest

18. 0.02, $-\frac{3}{5}$, -7, 4, 0.46, $-\frac{24}{6}$

   _____ _____ _____ _____ _____ _____
   least   greatest

Practice

19. _____ $= 34 * 78$

20. _____ $= 46 * 959$

21. $632 \div 4 =$ _____

22. $746 / 7 =$ _____

Name _____ Date _____ Time _____

STUDY LINK 11·7 | Capacity

Find at least one container that holds each of the amounts listed below.
Describe each container and record all the capacity measurements on the label.

1. Less than 1 Pint

Container	Capacity Measurements on Label
bottle of hot chili sesame oil	5 fl oz, 148 mL

2. 1 Pint

Container	Capacity Measurements on Label
bottle of cooking oil	16 fl oz, 473 mL

3. 1 Quart

Container	Capacity Measurements on Label

4. More than 1 Quart

Container	Capacity Measurements on Label

Complete.

5. 2 quarts = _____ pints

6. 3 gallons = _____ cups

7. _____ pints = 4 cups

8. _____ quarts = 12 cups

9. 6 pints = _____ quarts

10. _____ quarts = $2\frac{1}{2}$ gallons

Practice

11. $-3 + 7 =$ _____

12. _____ $= 3 + (-7)$

13. _____ $= 40 + (-80)$

14. $-60 + (-60) =$ _____

245

STUDY LINK 11·8

Unit 12: Family Letter

Rates

For the next two or three weeks, your child will be studying rates. Rates are among the most common applications of mathematics in daily life.

A rate is a comparison involving two different units. Familiar examples come from working (dollars per hour), driving (miles per hour), eating (calories per serving), reading (pages per day), and so on.

Our exploration of rates will begin with students collecting data on the rate at which their classmates blink their eyes. The class will try to answer the question "Does a person's eye-blinking rate depend on what the person is doing?"

During this unit, students will collect many examples of rates and might display them in a Rates All Around Museum. Then they will use these examples to make up rate problems, such as the following:

1. If cereal costs $2.98 per box, how much will 4 boxes cost?

2. If a car's gas mileage is about 20 miles per gallon, how far can the car travel on a full tank of gas (16 gallons)?

3. If I make $6.25 per hour, how long must I work to earn enough to buy shoes that cost $35?

Then the class will work together to develop strategies for solving rate problems.

The unit emphasizes the importance of mathematics to educated consumers. Your child will learn about unit-pricing labels on supermarket shelves and how to use these labels to decide which of two items is the better buy. Your child will see that comparing prices is only *part* of being an educated consumer. Other factors to consider include quality, the need for the product, and, perhaps, the product's effect on the environment.

This unit provides a great opportunity for your child to help with the family shopping. Have your child help you decide whether the largest size is really the best buy. Is an item that is on sale necessarily a better buy than a similar product that is not on sale?

Nutrition Facts	
Serving Size 1 link (45 g)	
Servings per Container 10	
Amount per Serving	
Calories 150 Calories from Fat 120	
	% Daily Value
Total Fat 13 g	**20%**
Total Carbohydrate 1 g	**<1%**
Protein 7 g	

Finally, students will look back on their experiences in the yearlong World Tour project and 50-facts test routine and share them with one another.

Please keep this Family Letter for reference as your child works through Unit 12.

Vocabulary

Important terms in Unit 12:

comparison shopping Comparing prices and collecting other information needed to make good decisions about which of several competing products or services to buy.

consumer A person who acquires products or uses services.

per *For each,* as in ten chairs per row or six tickets per family.

rate A comparison by division of two quantities with different units. For example, a speed such as 55 miles per hour is a rate that compares distance with time.

rate table A way of displaying *rate* information as in the miles per gallon table below.

Miles	35	70	105	140	175	210
Gallons	1	2	3	4	5	6

unit price The price *per* item or unit of measure. For example, if a 5-ounce package of something costs $2.50, then $0.50 per ounce is the unit price.

unit rate A *rate* with 1 in the denominator. For example, 600 calories per 3 servings or $\frac{600 \text{ calories}}{3 \text{ servings}}$ is not a unit rate, but 200 calories per serving $\left(\frac{200 \text{ calories}}{1 \text{ serving}}\right)$ is a unit rate.

Do-Anytime Activities

To work with your child on concepts taught in this unit, try these interesting and rewarding activities:

1. Have your child examine the Nutrition Facts labels on various cans and packages of food. The labels list the number of servings in the container and the number of calories per serving. Have your child use this information to calculate the total number of calories in the full container or package. *For example:*

   A can of soup has 2.5 servings.
   There are 80 calories per serving.
   So the full can has 2.5 * 80 = 200 calories.

2. Have your child point out rates in everyday situations. *For example:*

   store price rates: cost per dozen, cost per 6-pack, cost per ounce
   rent payments: dollars per month or dollars per year
   fuel efficiency: miles per gallon
   wages: dollars per hour
   sleep: hours per night
   telephone rates: cents per minute
   copy machine rates: copies per minute

3. Use supermarket visits to compare prices for different brands of an item and for different sizes of the same item. Have your child calculate unit prices and discuss best buys.

Building Skills through Games

In this unit, your child will play the following games. For more detailed instructions, see the *Student Reference Book.*

***Credits/Debits Game* (Advanced Version)** See *Student Reference Book*, page 239.
This game for 2 players simulates bookkeeping for a small business. A deck of number cards represents "credits" and "debits." Transactions are entered by the players on recording sheets. The game offers practice in addition and subtraction of positive and negative integers.

Fraction Top-It See *Student Reference Book*, page 247.
This game is for 2 to 4 players and requires one set of 32 Fraction Cards. The game develops skills in comparing fractions.

Name That Number See *Student Reference Book*, page 254.
This game is for 2 or 3 players and requires 1 complete deck of number cards. The game develops skills in representing numbers in different ways.

As You Help Your Child with Homework

As your child brings assignments home, you may want to go over the instructions together, clarifying them as necessary. The answers listed below will guide you through this unit's Study Links.

Study Link 12•1

2. $\frac{3}{5}$ **3.** $\frac{1}{8}$

4. 1   **5.** $\frac{5}{6}$

Study Link 12•2

1. $315

2. $12

3. 14 hours

4. a. 364 minutes per week

   b. 156 minutes

5. 9,096   **6.** 54,810

7. 81 R4   **8.** 13

Study Link 12•3

1. 2,100 feet

2. a. 3,500 pounds

   b. 420 gallons

3. 25 feet per second

4. a. 375 gallons

   b. 1,500 quarts

5. a. 480 feet

   b. 754 minutes, or $12\frac{1}{2}$ hours

6. 1,593   **7.** 55,080

8. 180 R4   **9.** 67

Study Link 12•4

1. 8 cents

2. $0.69

3. $0.35

4. Answers vary.

5. 1, 12; 2, 6; 3, 4

6. 1, 50; 2, 25; 5, 10

Study Link 12•5

1. $0.63

2. $0.37

3. $0.15

4. $0.35

5. $1.02

6. Sample answer: The 8-ounce cup is the better buy. The 8-ounce cup costs 9 cents per ounce, and the 6-ounce cup costs 10 cents per ounce.

7. Answers vary.

8. 1, 2, 3, 6, 7, 14, 21, 42

9. 1, 23

Study Link 12•6

1. 1,245 miles

2. About 9 times

3. a. About 69%

   b. About 49%

4. $\frac{8}{54}$, or $\frac{4}{27}$

5. a. China

   b. 6

   c. 9

   d. $9\frac{1}{2}$

Examples of Rates

1. Look for examples of rates in newspapers, in magazines, and on labels.

Study the two examples below, and then list some of the examples you find. If possible, bring your samples to class.

Example: _Label on a can of corn says "Servings Per Container 3½"_

Nutrition Facts
Serving Size 110 g
Servings Per Container 3 1/2

Amount Per Serving

Example: _Lightbulbs come in packages of 4 bulbs. The package doesn't say so, but there are always 4 bulbs in each package._

Example: _____

_____

_____

_____

Example: _____

_____

_____

_____

Example: _____

_____

_____

_____

Practice

2. $\frac{4}{5} - \frac{1}{5} =$ _____

3. _____ $= \frac{7}{8} - \frac{3}{4}$

4. _____ $= \frac{1}{9} + \frac{8}{9}$

5. $\frac{1}{3} + \frac{3}{6} =$ _____

STUDY LINK 12·2   **Rates**

Solve the problems.

1. Hotels R Us charges $45 per night for a single room.
   At that rate, how much does a single room cost *per week*? $_____

2. The Morales family spends about $84 each week for
   food. On average, how much do they spend *per day*? $_____

3. Sharon practices playing the piano the same amount
   of time each day. She practiced a total of 4 hours
   on Monday and Tuesday combined. At that rate,
   how many hours would she practice *in a week*?   _____ hours

Hours							
Days	1	2	3	4	5	6	7

Try This

4. People in the United States spend an average of 6 hours and 4 minutes
   each week reading newspapers.

   a. That's how many minutes *per week*?   _____ minutes per week

   b. At that rate, how much time does an average
      person spend reading newspapers in a *3-day period*?   _____ minutes

Minutes							
Days	1	2	3	4	5	6	7

Practice

5. _____ = 24 * 379   6. 870 * 63 = _____

7. 652 ÷ 8 = _____   8. 546 ÷ 42 = _____

STUDY LINK
12·3

Mammal Rates

1. A mole can dig a tunnel 300 feet long in one night.
   How far could a mole dig in one week?   About _____ feet

2. An elephant may eat 500 pounds of hay and drink
   60 gallons of water in one day.

   a. About how many pounds of hay could an
   elephant eat per week?   About _____ pounds

   b. About how many gallons of water could an
   elephant drink per week?   About _____ gallons

3. The bottle-nosed whale can dive to a depth
   of 3,000 feet in 2 minutes. About how many
   feet is that per second?   About _____ feet per second

4. A good milking cow will give up to 1,500 gallons of milk in a year.

   a. About how many gallons is that in 3 months?   About _____ gallons

   b. About how many *quarts* is that in 3 months?   About _____ quarts

---

Try This

5. Sloths spend up to 80 percent of their lives sleeping. Not only is a sloth extremely
   sleepy, but it is also very slow. A sloth travels on the ground at a speed of about
   7 feet per minute. In the trees, its speed is about 15 feet per minute.

   a. After one hour, how much farther would a sloth have
   traveled in the trees than on the ground (if it didn't
   stop to sleep)?   About _____ feet

   b. About how long would it take a sloth to travel 1 mile
   on the ground? (*Hint:* There are 5,280 feet in a mile.)   About _____ minutes,

                                                               or _____ hours

---

Practice

6. 59 * 27 = _____   7. _____ = 648 * 85

8. 904 ÷ 5 = _____   9. _____ = 536 / 8

255

STUDY LINK 12·4 | Unit Prices

Solve the unit price problems below. Complete the tables if it is helpful to do so.

1. A 12-oz bag of pretzels costs 96 cents. The unit price is _____ per ounce.

Dollars				0.96
Ounces	1	3	9	12

2. A package of 3 rolls of paper towels costs $2.07. The unit price is _____ per roll.

Dollars			2.07
Rolls	1	2	3

3. A 4-liter bottle of water costs $1.40. The unit price is _____ per liter.

Dollars				1.40
Liters	1	2	3	4

4. Choose 4 items from newspaper ads. In the table below, record the name, price, and quantity of each item. Leave the Unit Price column blank.

Item	Quantity	Price	Unit Price
Golden Sun Raisins	24 ounces	$2.99	

Practice

Name the factor pairs for each number.

5. 12 _____ **6.** 50 _____

257

STUDY LINK 12·5 | **Unit Pricing**

1. A package of 3 muffins costs $1.89.
   What is the price *per muffin?* _____

2. A 5-pound bag of rice costs $1.85.
   What is the price *per pound?* _____

3. Chewy worms are sold at $2.40 per pound.
   What is the price *per ounce?* _____

4. A 6-pack of bagels costs $2.11.
   What is the price *per bagel?* _____

5. A 2-pound bag of frozen corn costs $2.03.
   What is the price *per pound?* _____

6. A store sells yogurt in two sizes: The 8-ounce cup costs 72 cents, and the
   6-ounce cup costs 60 cents. Which is the better buy? Explain your answer.

   _____

   _____

   _____

   _____

7. Make up your own "better buy" problem. Then solve it.

   _____

   _____

   _____

   _____

   _____

Practice

Name all the factors.

8. 42 _____

9. 23 _____

STUDY LINK
12·6

Country Statistics

SRB
175 176

1. China has the longest border in the world—13,759 miles.
   Russia has the second longest border in the world—12,514 miles.
   How much shorter is Russia's border than China's border?   _____ miles

2. The area of Russia is about 1,818,629 square miles. The area of
   Spain, including offshore islands, is about 194,897 square miles.
   About how many times larger is Russia than Spain?   _____ times larger

3. Students in China attend school about 251 days per year.
   Students in the United States attend school about 180 days
   per year.

   a. About what percent of the year do Chinese students
      spend in school?   _____%

   b. About what percent of the year do American students
      spend in school?   _____%

4. English is officially spoken in 54 countries. Portuguese is
   officially spoken in 8 countries. Portuguese is spoken in about
   what fraction of the number of English-speaking countries?   _____

5. The table to the right shows
   the countries in the world with
   the most neighboring countries.

Country	Number of Neighbors
Brazil	10
China	15
Dem. Rep. of Congo	9
Germany	9
Russia	14
Sudan	9

   Use the data in the table to answer the following questions.

   a. Which country has the maximum number of neighbors? _____

   b. What is the range? _____

   c. What is the mode? _____

   d. What is the median? _____

Congratulations!

By completing *Fourth Grade Everyday Mathematics,* your child has accomplished a great deal. Thank you for all of your support.

This Family Letter is a resource to use throughout your child's vacation. It includes an extended list of Do-Anytime Activities, directions for games that can be played at home, a list of mathematics-related books to check out over vacation, and a sneak preview of what your child will be learning in *Fifth Grade Everyday Mathematics.* Enjoy the vacation!

Do-Anytime Activities

Mathematics means more when it is rooted in real-life situations. To help your child review many of the concepts he or she has learned in fourth grade, we suggest the following activities for you and your child to do together over vacation. These activities will help your child build on the skills he or she has learned this year and help prepare him or her for *Fifth Grade Everyday Mathematics.*

1. Have your child practice any multiplication and division facts that he or she has not yet mastered. Include some quick drills.

2. Provide items for your child to measure. Have your child use personal references, as well as U.S. customary and metric measuring tools.

3. Use newspapers and magazines as sources of numbers, graphs, and tables that your child may read and discuss.

4. Have your child practice multidigit multiplication and division using the algorithms that he or she is most comfortable with.

5. Ask your child to look at advertisements and find the sale prices of items using the original prices and rates of discount or find rates of discount using original prices and sale prices. Have your child use a calculator and calculate unit prices to determine best or better buys.

6. Continue the World Tour by reading about other countries.

STUDY LINK
12·7 | **Family Letter** *cont.*

Building Skills through Games

The following section lists rules for games that can be played at home. You will need a deck of number cards, which can be made from index cards or by modifying a regular deck of cards as follows:

A regular deck of playing cards includes 54 cards (52 regular cards plus 2 jokers). Use a permanent marker to mark some of the cards:

◆ Mark each of the four aces with the number 1.

◆ Mark each of the four queens with the number 0.

◆ Mark the four jacks and four kings with the numbers 11 through 18.

◆ Mark the two jokers with the numbers 19 and 20.

Beat the Calculator

Materials number cards 1–10 (4 of each); calculator

Players 3

Directions

1. One player is the "Caller," one is the "Calculator," and one is the "Brain."

2. Shuffle the deck of cards and place it facedown.

3. The Caller draws two cards from the number deck and asks for their product.

4. The Calculator solves the problem with a calculator. The Brain solves it without a calculator. The Caller decides who got the answer first.

5. The Caller continues to draw two cards at a time from the number deck and asks for their product.

6. Players trade roles every 10 turns or so.

Example: The Caller draws a 10 and 7 and calls out "10 times 7." The Brain and the Calculator solve the problem.

The Caller decides who got the answer first.

Variation 1: To practice extended multiplication facts, have the Caller draw two cards from the number deck and attach a 0 to either one of the factors or to both factors before asking for the product.

Example: If the Caller turns over a 4 and a 6, he or she may make up any one of the following problems:

4 * 60   40 * 6   40 * 60

Variation 2: Use a full set of number cards: 4 each of the numbers 1–10, and 1 each of the numbers 11–20.

Building Skills through Games

Name That Number

Materials   1 complete deck of number cards

Players   2 or 3

Object of the game To collect the most cards

Directions

1. Shuffle the cards and deal five cards to each player. Place the remaining cards number-side down. Turn over the top card and place it beside the deck. This is the **target number** for the round.

2. Players try to match the target number by adding, subtracting, multiplying, or dividing the numbers on as many of their cards as possible. A card may be used only once.

3. Players write their solutions on a sheet of paper or a slate. When players have written their best solutions:

   ◆ They set aside the cards they used to name the target number.

   ◆ Replace them by drawing new cards from the top of the deck.

   ◆ Put the old target number on the bottom of the deck.

   ◆ Turn over a new target number, and play another hand.

4. Play continues until there are not enough cards left to replace all of the players' cards. The player who sets aside more cards wins the game.

   Example: Target number: 16   A player's cards: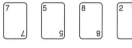

   Some possible solutions:

   $10 + 8 - 2 = 16$ (*three cards used*)

   $7 * 2 + 10 - 8 = 16$ (*four cards used*)

   $8 / 2 + 10 + 7 - 5 = 16$ (*all five cards used*)

   The player sets aside the cards used to make a solution and draws the same number of cards from the top of the deck.

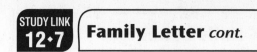

STUDY LINK 12·7   **Family Letter** *cont.*

Vacation Reading with a Mathematical Twist

Books can contribute to children's learning by presenting mathematics in a combination of real-world and imaginary contexts. The titles listed below were recommended by teachers who use *Everyday Mathematics* in their classrooms. They are organized by mathematical topic. Visit your local library and check out these mathematics-related books with your child.

Geometry

A Cloak for the Dreamer by Aileen Friedman

The Greedy Triangle by Marilyn Burns

Measurement

The Magic School Bus Inside the Earth by Joanna Cole

The Hundred Penny Box by Sharon Bell Mathis

Numeration

Alexander, Who Used to be Rich Last Sunday by Judith Viorst

If You Made a Million by David M. Schwartz

Fraction Action by Loreen Leedy

How Much Is a Million? by David M. Schwartz

Operations

Anno's Mysterious Multiplying Jar by Masaichiro Anno

The King's Chessboard by David Birch

One Hundred Hungry Ants by Elinor J. Pinczes

A Remainder of One by Elinor J. Pinczes

Patterns, Functions, and Sequences

Eight Hands Round by Ann Whitford Paul

Visual Magic by David Thomas

Reference Frames

The Magic School Bus: Inside the Human Body by Joanna Cole

Pigs on a Blanket by Amy Axelrod

Looking Ahead: Fifth Grade Everyday Mathematics

Next year your child will . . .

◆ Develop skills with decimals and percents

◆ Continue to practice multiplication and division skills, including operations with decimals

◆ Investigate methods for solving problems using mathematics in everyday situations

◆ Work with number lines, times, dates, and rates

◆ Collect, organize, describe, and interpret numerical data

◆ Further explore the properties, relationships, and measurement of 2- and 3-dimensional objects

◆ Read, write, and use whole numbers, fractions, decimals, percents, negative numbers, and exponential notation

◆ Explore scientific notation

Again, thank you for all of your support this year. Have fun continuing your child's mathematical experiences throughout the vacation!